praise for *all joy and no fun*

"If you are tempted to read just one more book on the arguably over-examined subject of parenthood, let it be Jennifer Senior's wise and surprising *All Joy and No Fun*."
—*Elle*

"*All Joy and No Fun* is chock-full of fascinating information from papers, studies, and books on wide-ranging subjects. . . . The thought-provoking nuggets it contains are valuable for any parent seeking perspective."
—Associated Press

"Neither a polemic nor a finger-wagger, [*All Joy and No Fun*] reveals a pastiche of perfectly lovely characters whose flaws and frailties and tiny triumphs make them all too human."
—*Chicago Tribune*

"[The] glimpses into the conundrums of other parents are thought-provoking and fun to read."
—*Newsday*

"[A] chatty, generous and yet statistically grounded reverse-angle of the usual studies of what parents do to children."
—*New York Post*

"If the rest of the book is superb, the final chapter of *All Joy and No Fun* is transcendent."
—*Brain, Child Magazine*

"A smart study of modern parenthood." —*Entertainment Weekly*

"Jennifer Senior can beautifully gloss a complicated academic text and then pull out a quote so lovely you want to tack it on your wall. . . . As the mother of a 3-year-old, I found myself underlining passages that begged to be shared with friends. . . . In short, *All Joy and No Fun* is a terrific read that speaks to something so present, yet so intangible: how each generation of children inevitably and irrevocably changes the generation of parents who bore them." —*BookPage*

"An indispensable map for a journey that most of us take without one. Brilliant, funny, and brimming with insight, this is an important book that every parent should read, and then read again. Jennifer Senior is surely one of the best writers on the planet."

—Daniel Gilbert, *New York Times* bestselling author of *Stumbling on Happiness*

"If you're a parent in the year 2014, you have to get your hands on a copy of this book. Wise, engrossing, and so real that I fear perhaps Jennifer Senior has been spying inside my house, *All Joy and No Fun* is a must-read for those of us whose lives have been immeasurably enriched and logistically derailed by having kids."

—Curtis Sittenfeld, bestselling author of *Prep* and *American Wife*

"A lovely, thoughtful book, written in a generous spirit and with a piercing intelligence. Jennifer Senior manages to mix unflinching social commentary with a warm and compassionate voice."

—Susan Cain, bestselling author of *Quiet: The Power of Introverts in a World That Can't Stop Talking*

"*All Joy and No Fun* captures the complex texture of parents' lives, the joys and the sorrows, highs and lows, with remarkable insight, intelligence, sensitivity, and subtlety."

—Alison Gopnik, bestselling author of *The Philosophical Baby*

"*All Joy and No Fun* is the perfect intellectual Rx for today's overstressed parents: A calm, clear-eyed synthesis of all the reasons their lives seem to be falling apart. The book's triumph is the way the author's own observations, presented with winning modesty and offhanded style, so brilliantly take down myths and assumptions to reveal why today's parents find their experience of raising children so different from what their own childhoods led them to expect. . . . This is a profound book about the meaning of love and how we raise not just our children but ourselves."

—Tom Reiss, author of *The Black Count*, winner of the 2013 Pulitzer Prize

"Traveling far beyond the infant and toddler years into the acute challenges of adolescence, Senior ingeniously deconstructs the kinds of experiences that all parents have but few parents talk about. By taking the reality of parenting out of the closet, she reveals in countless ways that none of us are in this alone. I loved this book."

—Madeline Levine, bestselling author of *Teach Your Children Well*

"Jennifer Senior has written a wonderful, smart, and deeply reported book that challenges many of the most sacred assumptions about modern parenthood. Written with authority and wisdom, it is destined to be the one book that all parents take with them on their mad, hair-raising, and, yes, joyous odyssey."

—David Grann, bestselling author of *The Lost City of Z*

PRAISE FOR *ALL JOY AND NO FUN*

"[An] astute book. . . . Refreshing. . . . Ms. Senior is a skilled interviewer, good at finding parents and children who defy expectations. . . . This is an eye-opening debut, and it will help a lot of parents feel less alone, if not less frazzled."
—JANET MASLIN, *NEW YORK TIMES*

"An important book, much the way *The Feminine Mystique* was, because it offers parents a common language, an understanding that they're not alone in their struggles, and an explanation of the cultural, political, and economic reasons for them."
—*CHRISTIAN SCIENCE MONITOR*

"Always generous in tone, Senior is a keen observer of the impact children have on their parents' marriages, mental health, work, and social lives, and she makes deft use of social-science research to tease out cultural shifts."
—*THE NEW YORKER*

"Richly woven, entertaining, enlightening, wrenching and funny."
—*WASHINGTON POST*

"Attention childless persons: If you're thinking of having kids, and are looking for an accurate assessment of the experience, disregard the holiday cards you may have received that portray merry families in various stages of triumph. Instead, read Jennifer Senior's book. This eloquent read is a tonic."
—HUFFINGTON POST

"Senior's wise compassion provides guidance that's both necessary and inspiring."
—*BOSTON GLOBE*

"Jennifer Senior's excellent new book . . . is not prescriptive. She doesn't tell parents to be more mindful or drink more wine or neglect their kids; she just wants them to understand why they are always so stressed out."
—HANNA ROSIN, SLATE

"A quick, lively read . . . [Senior's] carefully observed case studies of modern families read like scenes from novels. . . . She writes, 'The vocabulary for aggravation is large. The vocabulary for transcendence is more elusive.' She finds this vocabulary, inviting us into the heart-expanding moments of parenthood."
—*SAN FRANCISCO CHRONICLE*

all joy and no fun

all joy and no fun

THE PARADOX OF MODERN PARENTHOOD

jennifer senior

ecco

An Imprint of HarperCollinsPublishers

 For information address HarperCollins Publishers, 195 Broadway, New York, NY 10007.

HarperCollins books may be purchased for educational, business, or sales promotional use. For information please e-mail the Special Markets Department at SPsales@harpercollins.com.

A hardcover edition of this book was published in 2014 by Ecco, an imprint of HarperCollins Publishers.

FIRST ECCO PAPERBACK EDITION PUBLISHED IN 2015.

Designed by Suet Yee Chong

Library of Congress Cataloging-in-Publication Data has been applied for.

ISBN 978-0-06-207224-5

15 16 17 18 19 OV/RRD 10 9 8 7 6 5 4 3 2 1

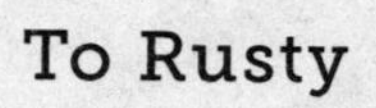

To Rusty

contents

introduction 1

one **autonomy** 15

two **marriage** 45

three **simple gifts** 95

four **concerted cultivation** 117

five **adolescence** 183

six **joy** 237

acknowledgments 267

a conversation with jennifer senior and curtis sittenfield 273

reader's guide 277

notes 281

index 309

all joy and no fun

introduction

THERE'S THE PARENTING LIFE of our fantasies, and there's the parenting life of our banal, on-the-ground realities. Right now, there's little question which one Angelina Holder is living. Eli, her three-year-old son, has just announced he's wet his shorts.

"Okay," says Angie, barely looking up. She's on a schedule, making Shake 'n Bake chicken parmesan for lunch. Her evening shift at the hospital starts at 3:00 P.M. "Go upstairs and change."

Eli is standing on a chair in the kitchen, picking at blackberries. "I can't."

"Why not?"

"I can't."

"I think you can. You're a big boy."

"I can't."

Angie unpeels the oven mitt from her hand. "What is Mommy doing?"

"Changing me."

"*No,* I'm cooking. So we're in a pickle."

Eli starts to whimper. Angie stops what she's doing. She looks annoyed, amused, and above all, baffled. There must be protocols for how to handle this kind of farcical exchange in parenting books, but she doesn't have time for books right now. She's got lunch to make, dishes to wash, and nursing scrubs to throw on.

"Why can't you change yourself?" she asks. "I want to hear this reasoning of yours."

"I can't."

Angie stares at her son. I can see her making the rapid calculation all parents make at this point in a cage match with a child, trying to determine whether it pays to relent. Eli is indeed capable of changing his own clothes, and unlike most three-year-olds, he usually succeeds on his first try, with his shirt facing forward and one limb in each pant leg. She could, in theory, hold her ground.

"Maybe you can go upstairs and get me new clothes for you to change into," she says, after mulling it over. "Maybe you can find me some green underwear. In your underwear bin?"

From an adult's perspective, this deal has all the face-saving elements of a good compromise. It's win-win. But Eli, being three, is not taking yes for an answer. Stalling, he wanders over to Angie's knapsack. "I think Zay wants this," he says, fishing out a granola bar. Zay, short for Xavier, is his younger brother.

"No, he doesn't." Angie is calm, but firm. She's picked a lane, and she's staying in it. "I need you to do what I ask. You're not listening right now."

Eli keeps sifting through the bag. Angie walks over and points him toward the stairs.

"I need help!" protests Eli.

"No, you don't," she answers. "I put all your clothes where they're supposed to be. Go upstairs and get them." A suspenseful couple of seconds tick by. Brinksmanship with a three-year-old. She looks conspiratorially at Zay. "Your brother's being silly, isn't he? What are we going to do with him?"

Eli huffs but capitulates, slowly making the climb to his room. A minute or so later, he appears at the top of the staircase, naked as Cupid, and tosses down a pair of clean green underwear.

"You *did* find your green underwear," Angie exclaims. "Good job!" She beams and pounces on it, as if it were a bridal bouquet.

BEFORE BECOMING A PARENT, Angie, it seems safe to say, would never have imagined that she'd be delighted to witness a preschooler throwing underwear down the stairs. She probably wouldn't have imagined the elaborate negotiation that preceded this gesture either, or that this kind of negotiation—at once ridiculous and agitating—would become a regular part of her mornings and afternoons. Before this, Angie worked as a psychiatric nurse in the evenings and biked and painted in her off-hours; on weekends, she went hiking with her husband at Minnehaha Falls. Her life was just her life.

But the truth is, there's little even the most organized people can do to prepare themselves for having children. They can buy all the books, observe friends and relations, review their own memories of childhood. But the distance between those proxy experiences and the real thing, ultimately, can be measured in light-years. Prospective parents have no clue what their children will be like; no clue what it will mean to have their hearts permanently annexed; no clue what it will feel like to second-guess so many seemingly simple decisions, or to be multitasking even while they're brushing their teeth, or to have a ticker tape of concerns forever whipping through their heads. Becoming a parent is one of the most sudden and dramatic changes in adult life.

In 1968, a sociologist named Alice Rossi published a paper that explored the abruptness of this transformation at great length. She called it, simply, "Transition to Parenthood." She noted that when it comes to having a child, there is no equivalent of courtship, which one does before marriage, or job training, which one does before, say, becoming a nurse. The baby simply appears, "fragile and mysterious" and "totally dependent."

At the time, it was a radical observation. In Rossi's day, scholars were mainly concerned with the effect of parents on their children. What Rossi thought to do was swing the telescope around and ask this question from the reverse perspective: What was the effect of parenthood on *adults*? How did having children affect their *mothers' and fathers'* lives? Forty-five years later, it's a question we're still trying to answer.

I FIRST STARTED THINKING about this question on the evening of January 3, 2008, when my son was born. But I didn't really explore it until more than two years later, when I wrote a story for *New York* magazine that examined one of the more peculiar findings in social science: that parents are no happier than nonparents, and in certain cases are considerably *less* happy.

This conclusion violates some of our deepest intuitions, but it stretches back nearly sixty years, even predating Rossi's research. The first report came in 1957, a peak time for the veneration of the nuclear family. The paper was called "Parenthood as Crisis," and in just four pages the author managed to destroy the prevailing orthodoxy, declaring that babies *weaken* marriages rather than save them. He quoted a representative mother: "We knew where babies came from, but we didn't know *what they were like* [emphasis his]." He then listed the complaints of the mothers he surveyed:

> Loss of sleep (especially during the early months); chronic "tiredness" or exhaustion; extensive confinement to the home and the resulting curtailment of their social contacts; giving up the satis-

> factions and the income of outside employment; additional washing and ironing; guilt at not being a "better" mother; the long hours and seven day (and night) week necessary in caring for an infant; decline in their housekeeping standards; worry over their appearance (increased weight after pregnancy, et cetera).

Fathers added more economic pressure, less sex, and "general disenchantment with the parental role" to the brew.

In 1975, another landmark paper showed that mothers presiding over an empty nest were not despairing, as conventional wisdom had always assumed, but *happier* than mothers who still had children at home; during the eighties, as women began their great rush into the workforce, sociologists generally concluded that while work was good for women's well-being, children tended to negate its positive effects. Throughout the next two decades, a more detailed picture emerged, with studies showing that children tended to compromise the psychological health of mothers more than fathers, and of single parents more than married parents.

Meanwhile, psychologists and economists started to stumble across similar results, often when they weren't looking for them. In 2004, five researchers, including the Nobel Prize–winning behavioral economist Daniel Kahneman, did a study showing which activities gave 909 working women in Texas the most pleasure. Child care ranked sixteenth out of nineteen—behind preparing food, behind watching TV, behind napping, behind shopping, behind *housework*. In an ongoing study, Matthew Killingsworth, a researcher at UC Berkeley and UC San Francisco, has found that children also rank low on the list of people whose company their parents enjoy. As he explained it to me in a phone conversation: "Interacting with your friends is better than interacting with your spouse, which is better than interacting with other relatives, which is better than interacting with acquaintances, which is better than interacting with parents, which is better than interacting with children. Who are on par with strangers."

These findings are undeniably provocative. But the story they tell is incomplete. When researchers attempt to measure parents' specific emotions, they get rather different—and much more nuanced—answers. Drawing from almost 1.8 million Gallup surveys collected between 2008 and 2012, researchers Angus Deaton and Arthur Stone found that parents with children at home experience more *highs,* as well as more lows, than those without children, a result that corresponds much more readily with our intuition. And when researchers bother to ask questions of a more existential nature, they find that parents report greater feelings of meaning and reward—which to many parents is what the entire shebang is about.

Children strain our everyday lives, in other words, but also deepen them. "All joy and no fun" is how a friend with two young kids described it.

Some people have flippantly concluded that these studies can be boiled down to one grim little sentence: *Children make you miserable.* But I think it's more accurate to call parenting, as the social scientist William Doherty does, "a high-cost/high-reward activity." And if the costs are high, one of the reasons may be that parenthood today is very different from what parenthood once was.

SOME OF THE HARDEST parts of parenting never change—like sleep deprivation, which, according to researchers at Queen's University in Ontario, can in some respects impair our judgment as much as being legally drunk. (There's something wonderfully vindicating about this analogy.) These perennial difficulties are worth dissecting and will certainly play a role in this book. But I am also interested in what's new and distinctive about modern parenting. There's no denying that our lives as mothers and fathers have grown much more complex, and we still don't have a new set of scripts to guide us through them. Normlessness is a very tricky thing. It almost guarantees some level of personal and cultural distress.

Obviously, there are hundreds of ways that the experience of parenting has changed in recent decades. But broadly speaking, I think three developments have complicated it more than most. The first is choice. Not all that long ago, mothers and fathers did not have the luxury of controlling how large their families were, or when each child arrived. Nor did they regard their children with the same reverence we modern parents do. Rather, they had children because it was customary, or because it was economically necessary, or because it was a moral obligation to family and community (often for all three reasons).

Today, however, adults often view children as one of life's crowning achievements, and they approach child-rearing with the same bold sense of independence and individuality that they would any other ambitious life project, spacing children apart according to their own needs and raising them according to their individual child-rearing philosophies. Indeed, many adults don't consider having children at all until they've deemed themselves good and ready: in 2008, 72 percent of college-educated women between the ages of twenty-five and twenty-nine had not yet had children.

Because so many of us are now avid volunteers for a project in which we were all once dutiful conscripts, we have heightened expectations of what children will do for us, regarding them as sources of existential fulfillment rather than as ordinary parts of our lives. It's the scarcity principle at work: we assign greater value to that which is rare—and those things for which we have worked harder. (In 2012, over 65,000 kids resulted from assisted reproductive technology.) As the developmental psychologist Jerome Kagan has written, so much meticulous family planning "inevitably endows the infant with a significance considerably greater than prevailed when parents had a half-dozen children, some at inauspicious times."

A popular but uncharitable way to interpret this change is to say that modern child-rearing has become a narcissistic undertaking. But there's a slightly more sympathetic way to think about this change too: by postponing children, many modern parents are far more aware of the freedoms they're giving up.

THERE'S A SECOND REASON our parenting experience has recently become more complicated: our *work* experience has gotten more complicated. We carry on with our day jobs long after we arrive home and kick off our shoes (the smart phone continues to ping; the home desktop continues to glow). Even more important, women's saturation of the labor market—the majority of mothers now work—has dramatically rewritten the rules of domestic life. In 1975, 34 percent of women with children under the age of three were in the workforce. In 2010, that number jumped to 61 percent.

That women bring home the bacon, fry it up, serve it for breakfast, and use its greasy remains to make candles for their children's science projects is hardly news. Yet how parenting responsibilities get sorted out under these conditions remains unresolved. Neither government nor private business has adapted to this reality, throwing the burden back onto individual families to cope. And while today's fathers are more engaged with their children than fathers in any previous generation, they're charting a blind course, navigating by trial and, just as critically, error. Many women can't tell whether they're supposed to be grateful for the help they're getting or enraged by the help they're failing to receive; many men, meanwhile, are struggling to adjust to the same work-life rope-a-dope as their wives, now that they too are expected to show up for Gymboree.

The result has been a lot of household aggravation. It's no accident that today's heirs to Erma Bombeck, the wicked satirist of domestic life who reigned in my mother's generation, are just as likely to be men as women. It was a man who wrote *Go the F**k to Sleep.* It was a male comic, Louis C.K., who developed a grateful cult following of moms and dads. "When my kids were younger, I used to avoid them," he said in a Father's Day riff in 2011. "You want to know why your father spent so long on the toilet? *Because he's not sure he wants to be a father.*"

TO MY MIND, THOUGH, there is a third development that has altered our parenting experience above all others, and that is the wholesale transformation of the child's role, both in the home and in society. Since the end of World War II, childhood has been completely redefined.

Today, we work hard to shield children from life's hardships. But throughout most of our country's history, we did not. Rather, kids worked. In the earliest days of our nation, they cared for their siblings or spent time in the fields; as the country industrialized, they worked in mines and textile mills, in factories and canneries, in street trades. Over time, reformers managed to outlaw child labor practices. Yet change was slow. It wasn't until our soldiers returned from World War II that childhood, as we now know it, began. The family economy was no longer built on a system of reciprocity, with parents sheltering and feeding their children, and children, in return, kicking something back into the family till. The relationship became asymmetrical. Children stopped working, and parents worked twice as hard. Children went from being our employees to our bosses.

The way most historians describe this transformation is to say that the child went from "useful" to "protected." But the sociologist Viviana Zelizer came up with a far more pungent phrase. She characterized the modern child as "economically worthless but emotionally priceless."

Today parents pour more capital—both emotional and literal—into their children than ever before, and they're spending longer, more concentrated hours with their children than they did when the workday ended at five o'clock and the majority of women still stayed home. Yet parents don't know what it is they're supposed to *do*, precisely, in their new jobs. "Parenting" may have become its own activity (its own profession, so to speak), but its goals are far from clear. Children are no longer economic assets, so the only way to balance the books is to assume they are *future* assets, which requires an awful lot of investment, not to mention faith. Because children are now deemed emotionally precious, today's parents are also charged with the *psychological* well-being of their sons and daughters, which on the face of it may seem like a laudable goal. But it's a murky one, and not necessarily realistic: building confidence in children is not the same as teaching them to read or to change a tire on your car.

THIS BOOK ATTEMPTS TO look at the experience of parenthood systematically, piece by piece, stage by stage, in order to articulate—and in some cases quantify—what today's parents find so challenging about their lives. To give but one example: that exasperating back-and-forth between Angie and Eli? Researchers have been examining that kind of exchange for more than forty years. In 1971, for instance, a trio from Harvard observed ninety mother-toddler pairs for five hours and found that on average, mothers gave a command, told their child no, or fielded a request (often "unreasonable" or "in a whining tone") every three minutes. Their children, in turn, obeyed on average only 60 percent of the time. This is not exactly a formula for perfect mental health.

There's a lot more research out there that helps to explain why modern parents feel as they do. What I've tried to do here is knit it all together, recruiting from a wide variety of sources. I've looked at surveys about sex and charts measuring sleep; books about attention and essays about distraction; histories of marriage and chronicles of childhood; and a wide range of inspired studies that document phenomena as varied as when teenagers fight most intensely with their parents (between eighth and tenth grade) and who feels the most work-life conflict (dads). I've then tried to show how all this material appears in the lives of real families, in their kitchens and bedrooms, during carpools and over homework hour, as they go about their daily business.

A FEW CAUTIONARY WORDS:

While it is my sincere hope that parents will read this book to better understand themselves—and, by extension, be easier on themselves—I make few promises about being able to provide any usable child-rearing advice. Tilt your head and stare long enough, and it's possible you'll make some out. But that is not my primary objective. This is not a book about children. It's a book about parents. *What to Expect When You're Expecting* may describe the changes that accompany pregnancy. But what changes should you expect when your children are three, or nine, or fifteen? What should you expect once your children are redirecting the course of your marriage, your job, your friendships, your aspirations, your internal sense of self?

One other crucial caveat: this book is about the middle class. Some of the families here may be struggling more than others, but all have to wrestle with difficult economic realities, whether they're social workers or shift workers, doctors or installers of security systems. I spend little time in the precincts of the elite, because their concerns aren't especially relatable (practically every child in this book goes to public school). But I also do not focus on the poor, because the concerns of poor parents *as parents* are impossible to view on their own. They are inextricable from the daily pressure to feed and house themselves and their children. As many have noted—perhaps most recently Judith Warner in *Perfect Madness*—poor parents deserve a different kind of book, and far more than one.

BECAUSE THE INDIVIDUAL STAGES of parenting don't much look like one another (the pandemonium of the toddler years feels very different from the frustrations and anxieties generated by adolescents), I've organized this book in a chronological fashion. Chapters 1 and 2 focus on the two things that undergo the most radical transformations once a child is born: our sense of autonomy, which gets entirely upended, and our marriages, whose rites and bylaws are suddenly undone. Chapter 3, on the other hand, concentrates on the unique pleasures that very young children can bring. Chapter 4 is about the middle parenting years—elementary school mostly—when parents feel immense pressure to prepare their children for an increasingly competitive world, thereby turning afternoons and weekends into a long procession of extracurricular activities. And chapter 5 concentrates on the adolescent years, whose effects on parents are wildly underdiscussed. We now shelter and care for our children for so long that they live with us through their own biological metamorphosis into adulthood. Yet precious little has been written about this awkward arrangement, a gap in the literature that's made doubly weird when one considers that parents, at this same moment, are going through significant life changes of their own, such as menopause and midcareer evaluations.

But my goal isn't just to analyze the difficulties of parenthood. The "high rewards," as William Doherty calls them, are worth analyzing too—they're just incredibly hard to measure. Meaning and joy have a way of slipping through the sieve of social science. The vocabulary for aggravation is large. The vocabulary for transcendence is more elusive. So in chapter 6, my last, I look at what raising children means in the larger context of a life—what it is to feel joy, what it is to surrender ourselves to a larger set of obligations, and what it is, simply, to tell our stories, to remember, to form whole visions of ourselves. We're all the sum of our experiences, and raising children plays an enormous part in making us who we are. For some of us, perhaps the largest part.

chapter one

autonomy

I held the baby up to the light, squinted at the physician out of one bloodshot eye, and spoke starkly: "Tell me, Doctor. You've been in this business a long time." I glanced meaningfully at the baby. "She's ruining my life. She's ruining my sleep, she's ruining my health, she's ruining my work, she's ruining my relationship with my wife, and . . . and she's ugly." . . . Swallowing hard, I managed to compose myself for my one simple question: "Why do I like her?"

—Melvin Konner, *The Tangled Wing* (1982)

WHEN I FIRST MET Jessie Thompson, it was mid-March, a trying time for Minnesota parents. Everywhere else in the country, spring had sprung; here, it would be at least another month before the kids could be humanely disgorged into the yard. All week long, I attended Early Childhood Family Education classes in and around Minneapolis and St. Paul, listening to roughly 125 parents talk about their lives. And all week long, at some point or another, almost all would give the same report: their nerves were shot and so were their kids' toys—the Play-Doh reduced to dry chips, the Legos scattered in a housewide diaspora. Everyone had the look of a passenger who'd been trapped far too long in coach and could not wait, for the love of everything that was holy, to deplane.

Minnesota's Early Childhood Family Education program (or ECFE, as I'll be referring to it from now on) is immensely popular and unique to the state, which is the reason I've come here. For a sliding-scale fee—and in some cases, no fee at all—any parent of a child who's not yet in kindergarten can attend a weekly class. And they do, in great numbers: in 2010, nearly 90,000 moms and dads signed up for one. The themes of the classes vary, but what they all have in common is an opportunity for parents to confide, learn, and let off steam.

The first half of each class is straightforward, with parents and children interacting in a group facilitated by ECFE's staff of early childhood education professionals. But the second half—*that's* when things get interesting. The parents leave their kids in the hands of those same professionals and retreat to a room of their own, where for sixty blissful minutes they become grown-ups again. Coffee is consumed; hair is let down; notes are compared. A parent educator always guides the discussion.

I met Jessie in one of the smaller ECFE classes in South Minneapolis and instantly liked her. She was one of those curious women who seemed not to realize she was pretty, carrying herself in a slightly distracted way. Her contributions to the discussion, while often wry ("I blame Oprah"), also suggested that she wasn't afraid of her darker, spikier feelings, and that she could even take a dispassionate view of them, as a lab researcher might of her rats. About midway through the class, for instance, she mentioned that she'd managed to get out of the house the previous evening to meet a girlfriend—a triumph, considering she had three kids under the age of six—"and I had this moment," she said, "where I realized, *This is how it feels when moms run away from their kids.* I could see why moms get in their cars and just . . . keep driving." She luxuriated for a few minutes in the high of being alone—just her on the open road, no children strapped into the car seats. "And then I had this actual *fantasy* for a few minutes," she said. "*What if I just keep driving?*"

She was not seriously entertaining this idea. Jessie was clearly a secure mother, which was why she was comfortable enough to confess this fleeting vision aloud. It was also clear, though, that she was dead-tired and not a little overwhelmed. She was trying to expand her new portrait photography business, based in the den of her home; she was living paycheck to paycheck; her youngest was just eight months old. She didn't have the resources to put her children in ballet classes and soccer, much less something as luxurious as preschool. She couldn't afford a babysitter for so much as one morning a week. Every trip to the grocery store involved loading all three kids in the car. "I just have these selfish bouts sometimes," she said. "Like: *I don't want to change another diaper. I don't want my kids hanging all over me 24/7. I want to have a phone conversation without being interrupted.*"

She was simply craving a few perks of her old life. But they were hard to come by with three small children in the house. Perhaps Erma Bombeck put it best more than thirty years ago when one of her characters declared: "I have not been alone in the bathroom since October."

ONE DAY YOU ARE a paragon of self-determination, coming and going as you please; the next, you are a parent, laden with gear and unhooked from the rhythms of normal adult life. It's not an accident that the early years of parenting often register in studies as the least happy ones. They're the bunker years, short in the scheme of things but often endless-seeming in real time. The autonomy that parents once took for granted has curtly deserted them, a fact that came up again and again among ECFE parents.

One father who'd opted to stay home with his two kids told his group—all stay-at-home dads—about running into a former colleague who was heading to Cuba for work. "And I was like, 'Wow, that's great,'" he said, gnashing his teeth, making it clear that he in fact thought it was the least great thing he'd heard in a while. He added:

> I see people who seem a lot more free, and they're doing things I wish that I could do, but for the fact that I have my family. Of course, did I want a family? Yes, I did. And do I get a lot of joy out of my children? Yes, I do. But in the day-to-day, it's sometimes hard to see. You rarely get a chance to do what you want, when you want.

Until fairly recently, what parents *wanted* was utterly beside the point. But we now live in an age when the map of our desires has gotten considerably larger, and we've been told it's our right (obligation, in fact) to try to fulfill them. In an end-of-the-millennium essay, the historian J. M. Roberts wrote: "The 20th century has spread as never before the idea that human happiness is realizable on Earth." That's a wonderful and energizing notion, certainly, but hardly always realistic, and when reality falls short of expectations, we often blame ourselves. "Our lives become an elegy to needs unmet and desires sacrificed, to possibilities refused, to roads not taken," writes the British psychoanalyst Adam Phillips in his 2012 collection of essays, *Missing Out*. "The myth of potential makes mourning and complaining feel like the realest things we ever do." Even if our dreams were never realizable, even if they were false from the start, we regret not pursuing them. "We can't imagine our lives," writes Phillips, "without the unlived lives they contain." And so we ask: *What if I just keep driving?*

Today's adults have an added reason to be spooked by those unlived lives: they have more time to exploit their potential before their children come along. Using National Vital Statistics birth data from 2010, a report by the National Marriage Project recently calculated that the average age of a college-educated woman at first birth is now 30.3 years old. The report added that college-educated women "typically have their first child more than two years after marrying." The consequence of this deferment is a heightened sense of contrast—before versus after. These parents now have an exquisite memory of what their lives were like before their children came along. They spent roughly a decade on their own, experimenting with different jobs, romantic partners, and living arrangements. That's twice as long as many of them spent in college.

During my week attending ECFE classes, few people talked about this before-and-after with more honesty or descriptive power than Jessie. In her early twenties she had taught English in Germany, worked at a pub in England, and done a brief stint as a flight attendant for Delta; now she was spending her days in a 1,700-square-foot bungalow with one bathroom (a lovely bungalow, but still). In her late twenties she had decided she wanted a career in advertising, and she was well en route to one by the time her first child was born; now she was presiding over a new, family-friendlier business (so she assumed), her peaceful downtown office replaced by a boisterous niche across from the TV room. "I really, really struggle with this still," she told her group. "It was just me and my husband until I was thirty-two."

Having children enlarges our lives in loads of unimaginable ways. But it also disrupts our autonomy in ways we couldn't have imagined, whether it's in our work, our leisure, or the banal routines of our day-to-day lives. So that's where this book begins: with a dissection of those reconfigured lives and an attempt to explain why they look and feel the way they do.

purloined sleep

One of the advantages to arriving at a household at 8:00 A.M.—assuming you can get past the inherent strangeness of everyone still half-clad in pajamas and walking around with uncombed hair—is that you can read in the parents' faces the story of both that morning and the evening before. When I visit Jessie in her South Minneapolis home a few months after our first meeting at ECFE, her husband, a civil engineer, is already long gone for work. But she's here and she's tired: it's clear that she either woke up early or went to bed late. It turns out the answer is both.

"Before you got here, I was so depressed," she confesses, shutting the door behind me. She's wearing a striped purple-and-maroon tank top, her long hair wet and bunched in a ponytail. Bella, five, and Abe, four, are both padding about, merry and oblivious to their mother's exhaustion, while the baby, William, is asleep upstairs. "The baby got up early," she explains. "And the others were up early too, and then the baby threw up on one of his stuffed animals." At roughly the same moment, Abe wet the bed, which meant the sheets had to be changed and he had to be bathed. Then William started spitting up juice in spectacular projectile fashion at breakfast. "This was at 7:37," she says. "I know, because I was thinking, *It's way too early for everything to be falling apart.*"

Which explains why she was up early. Why she was up late the night before is another story. Nighttime is Jessie's one opportunity for uninterrupted work, and she has an afternoon deadline today. Plus, she was fretting: she and her family will soon be decamping to the suburbs, in order to cut costs. The move should theoretically reduce her stress ("Half the taxes and half the price," she tells me), but she doesn't know a soul in her new community. Between her worries and her work, she didn't climb into bed until 3:00 A.M.

On some mornings, Jessie admits, she's so exhausted that the most she can do is set bowls of Cheerios and a cup of milk on the kitchen counter and then return to bed. "I do know a couple of moms who get enough sleep," she says. "I always wonder how they do it. Because I sure don't."

OF ALL THE TORMENTS of new parents, sleeplessness is the most infamous. But most parents-to-be, no matter how much they've been warned, don't fully grasp this idea until their first child comes along. Perhaps that's because they think they know what sleep deprivation feels like. But there's a profound difference between sustained sleep loss and the occasional bad night. David Dinges, one of the country's foremost experts on partial sleep deprivation, says that the population seems to divide roughly in thirds when it comes to prolonged sleep loss: those who handle it fairly well, those who sort of fall apart, and those who respond catastrophically. The problem is, most prospective parents have no clue which type they are until their kids come along. (Personally, I was the third type—just two bad nights, and *blam,* I was halfway down the loonytown freeway to hysterical exhaustion.)

Whatever type you may be—and Dinges suspects it's a fixed trait, evenly distributed between women and men—the emotional consequences of sleep loss are powerful enough to have earned their own analysis by Daniel Kahneman and his colleagues, the ones who looked at those 909 Texas women and found that they ranked time with their children lower than doing laundry. The women who'd had six hours of sleep or less were in a different *league* of unhappiness, almost, than those who'd had seven hours or more. The gap in their well-being was so extreme that it exceeded the gap between those who earned under $30,000 annually and those who earned over $90,000. (In newspapers and magazines, this finding is sometimes re-reported as "an hour extra of sleep is worth a $60,000 raise," which isn't exactly right, but close enough.)

A 2004 poll by the National Sleep Foundation found that parents of children two months old and younger slept, on average, just 6.2 hours during the night, and the numbers weren't much better for parents of children ten years old and younger, with their reports averaging only 6.8 hours of sleep per night. Other studies aren't quite so bleak: Hawley Montgomery-Downs, a neuroscientist who has done lots of work on this topic, recently found that parents of newborns average the same amount of sleep per night as nonparents—7.2 hours per night—with the crucial difference being that it's noncontinuous.

No matter which study you're consulting, though, most research suggests that the sleep patterns of new parents are fragmented, unpredictable, and just plain rotten, failing to do the one thing we cherish most about sleep, which is restore us to a sane state of body and mind. As I noted in the introduction, just a brief period of sleep deprivation compromises a person's performance as much as consuming excess alcohol. "So you can imagine the effects of sleeping for four hours every night for three months," says Michael H. Bonnet, a sleep researcher and clinical director of Kettering Medical Center in Dayton, Ohio. "We tend to think of it as a list of bad side effects: 'Well, this happens and this happens and this happens.' But it's the comparison with the alcohol studies that *really* makes the point, because we have agreed, as a society, that driving while drunk is punishable."

Bonnet adds that the sleep-deprived score higher on measures of irritability and lower on measures of inhibition too, which isn't an especially useful combination for parents, who are trying to keep their cool. Psychologists in fact have a term for the slow, incremental erosion of our self-restraint: they call it "ego depletion." In 2011 the psychologist Roy F. Baumeister and *New York Times* columnist John Tierney wrote a book on the subject called *Willpower,* whose central argument is that self-control, unfortunately, is not a bottomless resource. One of the most intriguing studies cited by the authors concluded, after following more than two hundred subjects throughout the day, that "the more willpower people expended, the more likely they became to yield to the next temptation that came along."

For me, this finding raises a question: assuming that parents spend a great deal of time fighting off the urge to sleep—and the urge to sleep is one of the two most common urges that adults try to fight (the other being the urge to eat)—then what urges do parents later succumb to instead? The most obvious answer I can think of is the urge to yell, an upsetting thought—nothing makes a mother or father feel quite so awful as hollering at the most vulnerable people in their lives. Yet that's what we do. Jessie confesses it's what she does, in spite of her enviably mellow disposition. "I'll yell," she says, "and then I'll feel bad that I yelled, and then I'll feel mad at myself: *Why didn't I get enough sleep?*"

pashas of excess

Five-year-old Bella wanders into the kitchen, where her mother and I have parked ourselves. Jessie cups her daughter's face in her hands. "What's up?"

"I'm hungry."

"So what do you say?"

"May I have something to eat, please?"

"Yes." Jessie flings open the fridge. Bella stares into it. Abe wanders over. The baby, William, is still down for his morning nap. "Abe, you want some yogurt?"

"Yeah."

"Yes please, Mama," Jessie corrects. "*You're the best, Mama.*" She smiles and rolls her eyes. Too much to ask for, obviously, but a woman can dream. "Arc you guys going to make apple pie?" Jessie's not talking about apple pie in the traditional sense, but something the kids invented: yogurt topped with applesauce and Cheerios and cinnamon. They have "pie-eating" contests sometimes, to see who got the ratios just right.

"Yes!"

The kids head out to the dining room while we remain in the kitchen. All is quiet for a little while. But a few minutes later, as we walk through the dining room to Jessie's office, we see Abe pick up a Play-Doh set and absently begin to deposit it onto a blob of yogurt. "Abe, no!" Jessie says, lunging quickly to avert a gloppy mess. Too late. "Everything *off* the table until I wipe it up, okay?" It's the first time I've heard tension creep into her voice all morning. She's so calm one almost forgets that life with small children is a long-running experiment in contained bedlam. She wipes the yogurt silhouette away, then stops for a brief second, staring at a constellation of Cheerios and crackers behind William's high chair, which he'd obviously been tossing behind him earlier that morning. Should she even bother cleaning it up? The kids are about to embark on another grubby project anyway, rolling Play-Doh hot dogs all over the dining room table. "Later," she decides, and continues into her office.

IN HIS 2005 COLLECTION of essays *Going Sane*, Adam Phillips makes a keen observation. "Babies may be sweet, babies may be beautiful, babies may be adored," he writes, "but they have all the characteristics that are identified as mad when they are found too brazenly in adults." He lists those characteristics: Babies are incontinent. They don't speak our language. They require constant monitoring to prevent self-harm. "They seem to live the excessively wishful lives," he notes, "of those who assume that they are the only person in the world." The same is true, Phillips goes on to argue, of young children, who want so much and possess so little self-control. “The modern child,” he observes. “Too much desire; too little organization.” Children are pashas of excess.

If you’ve spent most of your adult life in the company of other adults—especially in the workplace, where social niceties are observed and rational discourse is generally the coin of the realm—it requires some adjusting to spend so much time in the company of people who feel more than think. (When I first read Phillips’s observations about the parallels between children and madmen, it so happened that my son, three at the time, was screaming from his room, “I. Don’t. Want. To. Wear. PANTS.”)

Yet children do not see themselves as excessive. “Children would be very surprised,” Phillips writes, “to discover just how mad we think they are.” The real danger, in his view, is that children can drive *their parents* crazy. The extravagance of children’s wishes, behaviors, and energies all become a threat to their parents’ well-ordered lives. “All the modern prescriptive childrearing literature,” he concludes, “is about how not to drive someone (the child) mad and how not to be driven mad (by the child).”

This insight helps clarify why parents so often feel powerless around their young children, even though they’re putatively in charge. To a preschooler, all rumpus room calisthenics—whether it’s bouncing from couch cushion to couch cushion, banging on tables, or heaving bowls of spaghetti onto the floor—feel normal. But to adults, the child looks as though he or she has suddenly slipped into one of Maurice Sendak’s wolf suits. The grown-up response is to put a stop to the child’s mischief, because that’s the adult’s job, and that’s what civilized living is all about. Yet parents intuit, on some level, that children are *meant* to make messes, to be noisy, to test boundaries. “All parents at some time feel overwhelmed by their children; feel that their children ask more of them than they can provide,” writes Phillips in another essay. “One of the most difficult things about being a parent is that you have to bear the fact that you have to frustrate your child."

THERE ARE BIOLOGICAL UNDERPINNINGS that help explain why young children drive us crazy. Adults have a fully developed prefrontal cortex, the part of the brain that sits just behind the forehead, while the prefrontal cortexes of young children are barely developed at all. The prefrontal cortex controls executive function, which allows us to organize our thoughts and (as a result) our actions. Without this ability, we cannot focus our attention. And this, in some ways, is one of the most frustrating aspects of dealing with little kids: their attention is unfocused (or suffers from what Phillips might call "too little organization").

But again: children themselves do not perceive their attention as unfocused. In *The Philosophical Baby,* the psychologist and philosopher Alison Gopnik makes a distinction between a lantern and a spotlight: the spotlight illuminates just one thing while the lantern throws off a 360-degree glow. Adults have a spotlight consciousness. The consciousness of small children, on the other hand, is more like a lantern. By design, infants and preschoolers are highly distractible, like bugs with eyes all over their heads. And because the prefrontal cortex controls inhibitions as well as executive function, young children lack compunction about investigating every tangential object that captures their fancy. "Anyone who tries to persuade a three-year-old to get dressed for preschool will develop an appreciation of inhibition," she writes. "It would be so much easier if they didn't stop to explore every speck of dust on the floor."

You don't have to be especially clever to infer from this difference that adults might therefore find children a bit difficult to synchronize with their own agendas. A parent wants to put on a child's shoes and go to preschool; the child might agree, but then again, she might not, deciding it is vastly more important at that moment to play with her socks. Perhaps the parent has time to indulge this fascination, perhaps the parent doesn't. Either way, the parent must adapt, and that is hard: part of the reason we consider the world a comfortable place is because we can more or less predict the behavior of those in our lives. Small children send predictability out the window.

In addition to reason and focus and inhibition, the prefrontal cortex controls our ability to plan, to forecast, to ponder the future. But young children, whose prefrontal cortexes have barely begun to ripen, can't conceive of a future, which means they spend their lives in the permanent present, a forever feeling of *right now.* At times, this is a desirable state of consciousness; indeed, for meditators, it's the ultimate aspiration. But living in the permanent present is not a practical parenting strategy.

"Everybody would like to be in the present," says Daniel Gilbert, a social psychologist at Harvard and author of the 2006 best-seller *Stumbling on Happiness.* "Certainly it's true that there is an important role for being present in our lives. All the data say that. My own research says that." The difference is that children, by definition, *only* live in the present, which means that you, as a parent, don't get much of a chance. "Everyone is moving at the same speed toward the future," he says. "But your children are moving at that same speed with their eyes closed. So you're the ones who've got to steer." He thinks about this for a moment. "You know, back in the early seventies, I hung out with a lot of people who wanted to live in the present. And it meant that no one paid the rent."

In effect, parents and small children have two completely different temporal outlooks. Parents can project into the future; their young children, anchored in the present, have a much harder time of it. This difference can be a formula for heartbreak for a small child. Toddlers cannot appreciate, as an adult can, that when they're told to put their blocks away, they'll be able to resume playing with them at some later date. They do not care, when told they can't have another bag of potato chips, that life is long and teeming with potato chips. They want them *now,* because now is where they live.

Yet somehow mothers and fathers believe that if only they could convey the *logic* of their decisions, their young children would understand it. That's what their adult brains thrived on for all those years before their children came along: rational chitchat, in which motives were elucidated and careful analyses dutifully dispatched. But young children lead intensely emotional lives. Reasoned discussion does not have the same effect on them, and their brains are not yet optimized for it. "I do make the mistake of talking to my daughter sometimes like she's an adult," a woman named Kenya confessed to her ECFE group. "I expect her to understand. Like if I break things down enough, she'll get it."

The class instructor, Todd Kolod, nodded sympathetically. He'd heard it a thousand times before. It's the "little adult" problem, he explained. We mistakenly believe our children will be persuaded by our ways of reasoning. "But your three-year-old," he gently told her, "is never going to say, 'Yes, you're right. You have a point.' "

flow

"You want a dance party?" Jessie asks. "A pillow fight? A sword fight?" William has awakened from his morning nap, so she's taking a break from her work. One of the loveliest things about Jessie as a mother is that she seriously embraces play. She loves rocking out to music, loves art projects, loves clue games. (As in: "Something you pick." Answer: "Booger.") "Get off my boat!" she tells Abe, whose obsession of the moment is pirates. "Get on your *own* boat!" She picks up a light saber and jousts with one hand while cranking up the music on an iPod dock with the other. Then she picks up William, spins, and gives Abe a wicked look. "I'm stealing your boat! I'm going to take all your treasure!"

Abe bangs his light saber on the ground.

She looks mildly cross for a second. "Don't do that. You'll break it." Then, back in character: "Less talk, more action!" She leans in with her light saber to attack Abe, then gives it to William to do the same. She puts William down and begins to tickle Abe, who likes it at first, but objects when she moves in to devour his belly.

"Don't do that," he tells her. Their rhythm, again, is disrupted.

"Don't do that?" she says. "You know why I do that? Because I love you." She turns him upside down.

"No!" he repeats.

She looks at him assessingly. "You were up too early, huh? Okay. No swinging." She decides to change both songs and tactics, turning her son right side up and holding him in a koala hug as she finds a beautiful Spanish ballad. They start to slow-dance. It clicks. The music forms a cocoon around them, as if I'm not even there. Abe melts into his mother's shoulder. She breathes him in.

NO GRAPH IN THE world can do full justice to these unexpected moments. They're sweet little bursts of grace, and they leave sense-memories on the skin (the smell of the child's shampoo, the smoothness of his arms). That's why we're here, leading this life, isn't it? To know this kind of enchantment?

The question is why such moments, at least with small children, often feel so hard-won, so *shatterable,* and so fleeting, as if located between parentheses. After just a few minutes of this dreamy slow-dance with Abe, William does a face-plant and starts howling. Jessie sambas over and handles it with humor. This is the drill.

I'd like to propose a possible explanation for why these moments of grace are so rare: the early years of family life don't offer up many activities that lend themselves to what psychologists call "flow." Simply put, flow is a state of being in which we are so engrossed in the task at hand—so fortified by our own sense of agency, of *mastery*—that we lose all sense of our surroundings, as though time has stopped. Athletes commonly experience this feeling when they're sinking every shot or completing every pass ("being in the zone," they call it); artists commonly experience it too, when music or paint pours out of them as if they were mere spigots.

The paradoxical thing about flow is that it is often marked by an *absence* of feeling, experienced nonetheless as a form of undiluted bliss. That's what makes flow one of the most beguiling and equal-opportunity parts of our emotional lives: no matter what kind of temperament we've been handed, even if it's melancholic, almost all of us have the ability to lose ourselves in something we love and do well.

In order to experience this kind of magical engagement, though, circumstances need to align. This is where the work of the Hungarian psychologist Mihaly Csikszentmihalyi is a revelation. For decades, Csikszentmihalyi (pronounced as "cheeks sent me high") has been thinking about flow, analyzing the conditions that make it possible, and looking in broad cultural terms at what gives us our deepest satisfactions. He has dissected the flow experiences of thousands of people. In 1983, he even codeveloped an innovative technique to measure it, by contacting study participants at random intervals and asking them to record not just *what* they were doing at that moment but how they *felt* about it. (Bored? Engaged? In control? Scared? Stressed? Exhilarated?) He called this tool the Experience Sampling Method, or ESM. It was an inspired contribution to his field. Researchers for the first time were making the distinction between how study participants felt *in the moment* and how they felt retrospectively.

Eventually, Csikszentmihalyi began to notice common patterns in flow experience. Most flow experiences occur, for example, during situations that are "goal-oriented and bounded by rules." In fact, most activities that lend themselves to flow are designed to maximally engage our attention and expand our competence—like athletics or intense work. "They have rules that require the learning of skills," he writes in *Flow,* his 1990 book on the subject. "They set up goals, they provide feedback, they make control possible."

In theory, young children like rules. But they're pretty spotty observers of them. Every parent has a story about a perfectly planned day—a trip to the zoo, a jaunt to the local ice cream joint—that devolved into something close to anarchy. Most of life with young children does not have a script, and if a parent attempts to write one, children may not be inclined to follow it. That's what it means to look after people with immature prefrontal cortexes. Their neurocircuitry conspires against focus. Gopnik says it outright, midway through *The Philosophical Baby:* "This expansive lantern consciousness is almost the opposite of the distinctive adult happiness that comes with what psychologists call 'flow.'" To be in flow, one must pay close and focused attention. Yet very young children are wired for discovery, for sweeping in lots of stimuli. And if they can't be in flow, chances are you'll have a hard time slipping into flow yourself—in the same way that athletes have a much harder time finding their groove if their teammates are distracted.

This subject came up repeatedly in ECFE classes. At one point, Annette Gagliardi, a veteran instructor, started to ask the parents in one of her seminars whether having a focused plan for the day made them happier. One mother cut her off. "Only if the plan goes *well.* If there are meltdowns, it's *What was I thinking?*"

"Which is why I have very low expectations," said another. "You shoot for the bare minimum and are excited by anything else."

A clear plan isn't the only requirement for flow. Csikszentmihalyi also noticed that we enjoy ourselves most when we're positioned "at the boundary between boredom and anxiety, when the challenges are just balanced with [our] capacity to act." Yet parents of young children often describe the sensation of lurching back and forth *between* those two poles—boredom and anxiety—rather than being able to comfortably settle somewhere in the middle. "To the extent that we are not maximally happy when we're with our young children," says Daniel Gilbert, the social psychologist, "it could be that they're demanding things of us we find difficult to give. But it could also be that they're not demanding *that much*."

Consider what happens at the end of Jessie's impromptu dance party. Once William begins to wail, she has a hard time figuring out how, precisely, to console him. She tries rocking him, she tries giving him Cheerios; at one point, she even tries picking him up, while Abe is still on her shoulder. But the only thing that seems to work, in the end, is the simplest repeated act: tossing a pair of pants from the laundry basket over his head and yanking them off. "Where's William?" she asks. *Whoosh.* "There he is!" Another toss. "Where's William?" *Whoosh.* "There he is!" It's boring, sure, and there's certainly no flow. But it works.

Boredom can be an awkward topic for parents. It feels like a betrayal to admit that time spent with one's children isn't always stimulating. But even Benjamin Spock, the cuddly pediatrician who dominated the child-rearing advice market for the second half of the twentieth century, talked about it. "The fact is," he once wrote, "setting aside a chunk of time to be devoted exclusively to companionship with children is a somewhat boring prospect to a lot of good parents." Boredom also came up in the ECFE classes I attended, including Jessie's, with the instructor herself confessing that she found it dull to play "My Little Pony" when her daughter was small. "That was the most negative emotion I experienced as a father," recalls Gilbert. "Boredom.

Throwing the ball back and forth and back and forth and back and forth. The endless repetition, the can-you-do-it-again, the can-you-read-the-same-story-one-more-time. There were times I just thought, *Give me a gun.*"

In *Flow,* Csikszentmihalyi explains that most flow experiences happen *apart* from everyday life rather than in the midst of it. But raising children *is* everyday life. In Csikszentmihalyi's view, people have more control in specialized settings, even dangerous ones; hang-gliders, deep-sea divers, or race car drivers, writes Csikszentmihalyi, still "report flow experiences in which a heightened sense of control plays an important part," because they feel the *possibility* of success. Above all, people report experiences of flow while they're working. It sounds counterintuitive, but not if one considers how propitious work conditions are to flow: work provides rules, clear-cut goals, and immediate feedback.

After finishing *Flow,* the reader comes away with the unmistakable impression that most people find themselves in flow when they're alone. Csikszentmihalyi talks about fishing, cycling, and rock climbing; about solving equations, playing music, and writing poems. As a rule, the experiences he describes do not involve much social interaction, least of all with children.

I was so struck by *Flow*'s negative implications for parents that I decided I wanted to speak to Csikszentmihalyi, just to make sure I wasn't misreading him. And eventually I did, at a conference in Philadelphia where he was one of the marquee speakers. As we sat down to chat, the first thing I asked was why he talked so little about family life in *Flow.* He devotes only ten pages to it. "Let me tell you a couple of things that may be relevant to you," he said. And then he told a personal story. When Csikszentmihalyi first developed the Experience Sampling Method, one of the first people he tried it out on was himself. "And at the end of the week," he said, "I looked at my responses, and one thing that suddenly was very strange to me was that every time I was with my two sons, my moods were always very, very negative." His sons weren't toddlers at that point either. They were older. "And I said, 'This doesn't make any sense to me, because I'm very proud of them, and we have a good relationship.'" But then he started to look at what, specifically, he was doing with his sons that made his feelings so negative. "And what was I doing?" he asked. "I was saying, 'It's time to get up, or you will be late for school.' Or, 'You haven't put away your cereal dish from breakfast.'" He was nagging, in other words, and nagging is not a flow activity. "I realized," he said, "that being a parent consists, in large part, of correcting the growth pattern of a person who is not necessarily ready to live in a civilized society."

I asked if, in that same data set, he had any numbers about flow in family life. None were in his book. He said he did. "They were low. Family life is organized in a way that flow is very difficult to achieve, because we assume that family life is supposed to relax us and to make us happy. But instead of being happy, people get bored." Or enervated, as he'd said before, when talking about disciplining his sons. And because children are constantly changing, the "rules" of handling them change too, which can further confound a family's ability to flow. "And then we get into these spirals of conflict and so forth," he continued. "That's why I'm saying it's easier to get into flow at work. Work is more structured. It's structured more like a game. It has clear goals, you get feedback, you know what has to be done, there are limits." He thought about this. "Partly, the lack of structure in family life, which seems to give people freedom, is actually a kind of an impediment."

divided attention

It's early afternoon, William is down for his second nap, and Jessie is sitting in front of her computer, staring at an image from her most recent photo shoot. It's pretty wonderful—a woman pulling two kids in a red wagon—but Jessie's not pleased with it, and this client is scheduled to come by tomorrow evening. Jessie is determined to get her portfolio right.

Bella walks in. "Mom, I need help."

Jessie is still staring at the screen. "What's going on?"

"I want to do Roku."

"You can't do Roku right now. Watch a movie."

"I need *you*."

She sighs, gets up from her desk, and walks into the TV room, just opposite her office. "Bella, you need to change the channel. Here." She punches a button.

Flow is hard enough to achieve if your sole task is trying to care for your kids. But it's even harder if you're trying to care for your children *and* work at the same time. Today, that's what many of us are doing. According to the Bureau of Labor Statistics, roughly one-quarter of employed men and women work from home at least some of the time. Even those who work exclusively outside the home now find that the border between their living room and their workplace has dissolved. Once upon a time, only doctors had to live with after-hours disruptions. Now, many professionals walk around with the impression that everything they do is urgent. Emergencies are regular occurrences; late-night texts in all caps go with the territory. The portability and accessibility of our work has created the impression that we should always be available. It's as if we're all leading lives of *anti*-flow, of chronic interruptions and ceaseless multitasking.

This subject, too, surfaced and resurfaced in ECFE classes. Responding to the beckoning smart phone and the siren call of email—these turned out to be huge and surprisingly shameful refrains among parents, as if *their children* were the disruptions, rather than the other way around. One father summed up his feelings in two sentences:

"There are days I'm able to put work behind me and just *be* with my son, and it feels awesome. But then there are days when all I'm thinking is, *If I can get this kid taken care of, I can get back on the computer,* and it feels terrible."

Parents attempting to work out of their homes brought up this topic the most. Jessie talked at length about her divided attention—how difficult she found it, both emotionally and intellectually, to toggle between her portrait business and her children's needs. She knew she wanted to stay at home. Her own mother had died just two years before Bella came along, and the black abruptness of it crystallized, in her mind, the importance of being around as a parent. But she also came from a long line of female breadwinners, "women with master's degrees and women who ran companies." Anyway, she liked her work. It gave her a sense of independence and pride. But she couldn't figure out how to manage the rhythms and demands of both her family and her work at the same time, especially after William, her third, was born. "I think back to yesterday," she told her class, "and I knew what the good parent should do. I knew I should *stop*." She'd been editing a photo shoot, just as she is doing today, and William had started to cry. "I knew that if I gave him the bottle and I held him and I kissed him, it would be all right," she continued. "But I had this deadline over my head, and for some reason I couldn't let it go. So I'm emailing the parent, and I'm trying to work . . . all while feeling bad about myself and this choice. I'm not even sure why I made it. No one benefited in the end." You could see the confusion in her face.

Neurologically speaking, though, there are reasons we develop a confused sense of priorities when we're in front of our computer screens. For one thing, email comes at unpredictable intervals, which, as B. F. Skinner famously showed with rats seeking pellets, is the most seductive and habit-forming reward pattern to the mammalian brain. (Think about it: would slot machines be half as thrilling if you knew when, and how often, you were going to get three cherries?) Jessie would later say as much to me when I asked her why she was "obsessed"—her word—with her email: "It's like fishing. You just never know what you're going to get."

More to the point, our nervous systems can become dysregulated when we sit in front of a screen. This, at least, is the theory of Linda Stone, formerly a researcher and senior executive at Microsoft Corporation. She notes that we often hold our breath or breathe shallowly when we're working at our computers. She calls this phenomenon "email apnea" or "screen apnea." "The result," writes Stone in an email, "is a stress response. We become more agitated and impulsive than we'd ordinarily be."

One could still make the case that smart phones and living room WiFi have been a boon to today's middle-class parents, because they allow mothers and fathers the flexibility to work from home. The difficulty, in the words of Dalton Conley, an NYU sociologist, is that they allow "many professionals with children to work from home *all the time*." The result, he writes in his book *Elsewhere USA*, is that "work becomes the engine and the person the caboose, despite all this so-called freedom and efficiency." A wired home lulls us into the belief that maintaining our old work habits while caring for our children is still possible.

The problems with this arrangement are obvious. As Jessie observed, trying to do two things at once doesn't work so well. Humans may pride themselves on their ability to swing from one task to another and then back again, but task-switching isn't really a specialty of our species, as reams of studies have shown. According to Mary Czerwinski, another attention expert at Microsoft Corporation, we don't process information as thoroughly when we task-switch, which means that information doesn't sink into our long-term memories as deeply or spur us toward our most intelligent choices and associations. We also lose time whenever we switch tasks, because it takes a while to intellectually relax into a project and build a head of steam.

And that's just at the office. It's likely that our work suffers even more acutely when we're attempting to do it from home. Disruptions at the office—say, an email from a colleague inquiring about a memo—usually generate little emotional heat. Disruptions from children, on the other hand, often generate plenty of it, and strong emotions aren't easy to subdue. "There's a warm-up period," explains David E. Meyer, an expert on multitasking at the University of Michigan. "And then there's a calming-down period that happens subsequently. And both take extra time away from getting a task done. The hormones that happen after an emotion linger in the bloodstream for hours, sometimes days." Especially if the emotion is a negative one. "If the interlude involves anger or sadness," he says, "or the kinds of emotions Buddhists refer to as 'destructive,' they're going to have a much more negative impact on something you were doing that was emotionally neutral."

So imagine your child is having a meltdown while you're working. Or he's hungry, or skins his knee, or is fighting with his sister. We *physically* experience these disruptions differently. "This is over and above the stuff that happens when you switch between two different windows that are neutral in nature," says Meyer. "This is emotional task-switching. I don't know if anyone's ever used that term, but it has an additional layer to it."

The result, almost no matter where you cut this deck, is guilt. Guilt over neglecting the children. Guilt over neglecting work. Working parents feel plenty of guilt as it is. But in the wired age, to paraphrase Dalton Conley, parents are able to feel that guilt *all the time*. There's always something they're neglecting.

I am now watching this conflict unspool in real time in Jessie's home office. About thirty minutes after she's helped Bella set up the movie, Bella walks back in. "Mom? It's not doing that *brrrrrrrrrrrrrr—*" She rolls her *r*'s, attempting to imitate the sound of a whirring videotape. They still use a VCR.

"It's not rewinding?"

"No, it's not rewinding, and I want to watch Barney again."

Jessie gets up from her desk and goes to the family room with Bella, giving her a brief tutorial on how to rewind the tape. Then, for a third time, she returns to her office and tries to focus on her work, adjusting the light on an image that won't cooperate. She still hates it. "I'm afraid this looks over-Photoshopped."

Bella comes back through the door, this time with tears in her eyes. "It's still not working!"

Jessie looks intently at her daughter. "Is it worth crying over?" Her daughter, wearing a denim skirt with two hearts on the rear pockets, seems to consider this question. "Take a breath. A breath, please. Okay? Calm down." Jessie walks into the TV room. "See this?" She points to the VCR and then looks at Bella. "This button makes it go back to the beginning. And then you press Play."

She goes back to her office a fourth time and takes her seat. She has not spent more than thirty consecutive minutes in front of her computer since she started, and her husband won't be home until dinner. "Sometimes I notice that when the kids are really overwhelming me, work is a *big* release," she says. "But at this moment, I'm not trying to get away. I have a real deadline." She looks up. "I think I hear a baby." She does. William's awake. "Crap. I haven't gotten enough done." She fiddles with an image onscreen. "This job is very mental," she says. "When I'm doing a shoot, I'm thinking about light and background and wardrobe and props. When I'm editing, I'm trying to make the pictures look magical without looking over-Photoshopped." But then she gets lost in what she's doing, and the kids start to beckon. Like now. A few minutes go by. "See?" she looks up at me, waiting for me to notice what she's noticing. I give her a look indicating that I don't. "I keep telling myself, *I just want to edit this set I have open in Photoshop, and then I'll get William.*" She points upstairs. It's dead quiet. What she's noticed is an absence. We were both so absorbed in the photographs that we didn't realize William had stopped crying.

missing out?

Jessie could defer her professional dreams until her children are older. It's a trade-off plenty of women make. She could forgo the money, forgo the satisfaction. In so doing, she could at least find relief in consolidating her time and energy into one main project—her kids—and focus on that alone, rather than feel dogged all the time by a sense of guilt.

Or Jessie could make a different choice: she could scale up her business and get out of the house entirely. If she's going to contribute to the family economy and realize her full professional capabilities, she may as well go all out, right? Then, during work hours, she's doing work. Not rewinding Barney, not mopping yogurt off the dining room table. Of course, it's a costly proposition and may simply not be feasible: she'd have to take out a loan to make her business bigger. But it would afford her a better chance to experience flow. She'd be a photographer at work and a mother at home. Sure, the smart phone would still chirp and the inbox would still brim. But at least she'd have a formal division in place.

Jessie has instead chosen the hardest path. She's trying to do both, improvising all day long as she juggles her dual responsibilities, never knowing when her kids will require attention or when a work deadline will crop up.

It's a heady question, how women balance these concerns. Recently, the question has found its way back to the center of a contentious and very emotional debate. If you're Sheryl Sandberg, the chief operating officer of Facebook and author of *Lean In,* you believe that women should stop getting in their own way as they pursue their professional dreams—they should speak up, assert themselves, defend their right to dominate the boardroom and proudly wear the pants. If you're Anne-Marie Slaughter, the former top State Department official who wrote a much-discussed story about work-life balance for *The Atlantic* in June 2012, you believe that the world, as it is currently structured, cannot accommodate the needs of women who are ambitious in both their professions and their home lives—social and economic change is required.

There's truth to both arguments. They're hardly mutually exclusive. Yet this question tends to get framed, rather tiresomely, as one of how and whether women can "have it all," when the fact of the matter is that most women—and men, for that matter—are simply trying to keep body and soul together. The phrase "having it all" has little to do with what women want. If anything, it's a reflection of a widespread and misplaced cultural belief, shared by men and women alike: that we, as middle-class Americans, have been given infinite promise, and it's our obligation to exploit every ounce of it. "Having it all" is the phrase of a culture that, as Adam Phillips implies in *Missing Out,* is tyrannized by the idea of its own potential.

JUST A FEW GENERATIONS ago, most people didn't wake up in the morning and fret about whether or not they were living their lives to the fullest. Freedom has always been built into the American experiment, of course, but the freedom to take off and go rock-climbing for the afternoon, or to study engineering, or even to sneak in ten minutes for ourselves in the morning to read the paper—these kinds of freedoms were not, until very recently, built into our private universes of anticipation. It's important to remember that. If most of us don't know what to do with our abundant choices and the pressures we feel to make the most of them, it may simply be because they're so new.

The sociologist Andrew Cherlin makes this quite clear in his very readable 2009 book *The Marriage-Go-Round.* In the New England colonies, he notes, individual family members hardly expected time to themselves to pursue their own interests. There were too many children running around to allow anyone much peace and quiet (families in Plymouth averaged seven or eight kids each), and the architecture of the typical Puritan home conspired against solitary endeavors, with most activities concentrated in one main room. "Personal privacy," he writes, "one of the taken-for-granted aspects of modern individualism, was in short supply." From the moment of birth, people were enmeshed in a complex web of obligations and formal roles, and throughout their lives, they were expected to follow scripts that helped fulfill them.

It wasn't until industrialization—and by extension, urbanization—that people began to have more control over their fates. For the first time, droves of young men left the orbit of their homes to find work in the factories of the expanding cities, meaning that they got to choose *both* their vocations and their wives. Women, too, gained a bit more control over their lives as the twentieth century progressed. People are often surprised to hear this, assuming that women had no agency at all until the late 1960s, with the blooming of the women's movement. But in *The Way We Never Were,* the historian Stephanie Coontz shows that women worked outside the home steadily, and in increasing numbers, throughout the twentieth century. The 1950s, putatively the golden age of the family, were the real anomaly: the median age of women at first marriage fell to twenty (in 1940, it was twenty-three); birth rates increased (the number of women with three or more children doubled over twenty years); and women started dropping out of college at a much faster rate than men.

But by the 1960s, the college dropout rate between the sexes had evened out again, better positioning women for more opportunities in the workplace. The 1960s also brought the Pill, which gave women unprecedented freedom to plan their families (and choose their husbands, for that matter, by allowing them to avoid marriages forced by unwanted pregnancies). Then came the more liberal divorce laws of the 1970s, allowing women the economic freedom to leave marriages that made them unhappy.

The culmination of all these developments was a culture abundant in choice, with middle-class American men and women at liberty to chart the course of their lives in all sorts of ways that historically had been unthinkable. And the liberalization of the 1970s was nothing compared to today's emphasis on self-realization. "Regardless of their educational level, Americans face a situation in which lifestyle choices, which were limited and optional a half century ago, are now mandatory," writes Cherlin. "You *must* [emphasis his] choose, again and again. The result is an ongoing self-appraisal of how your personal life is going, like having a continual readout of your emotional heart rate."

Few of us would want to reverse the historical advances that gave us our newfound freedoms. They're the hard-won products of economic prosperity, technological progress, and the expansion of women's rights. My mother had to marry at twenty in order to get out of her parents' house and into a world of her own. The triumph of the women of her generation was to rewrite this rule—"get un-married and be free," as Claire Dederer puts it in her beautiful memoir, *Poser*—which made it possible for their daughters to rent apartments, settle into careers, marry later, and even leave those marriages if they didn't work out.

But these gains in freedom for both men and women often seem like a triumph of subtraction rather than addition. Over time, writes Coontz, Americans have come to define liberty "negatively, as lack of dependence, the right not to be obligated to others. Independence came to mean immunity from social claims on one's wealth or time."

If this is how you conceive of liberty—as freedom *from* obligation—then the transition to parenthood is a dizzying shock. Most Americans are free to choose or change spouses, and the middle class has at least a modicum of freedom to choose or change careers. But we can never choose or change our children. They are the last binding obligation in a culture that asks for almost no other permanent commitments at all.

Which leads back to Jessie's fantasy of getting in the car, pulling onto the highway, and continuing to drive. She can't, naturally, and never would. That itinerary exists only in her mind. No matter how perfect our circumstances, most of us, as Adam Phillips observes, "learn to live somewhere between the lives we have and the lives we would like." The hard part is to make peace with that misty zone and to recognize that no life—no life worth living anyway—is free of constraints.

chapter two

marriage

> *My wife's anger toward me seemed barely contained. "You only think about yourself," she would tell me. "I never thought I'd have to raise a family alone."* —Barack Obama, *The Audacity of Hope* (2006)

JESSIE THOMPSON'S ECFE CLASS was small and intense. Angelina Holder's, on the other hand, was big and raucous. The women spoke with the ease of those who'd already heard each other's life stories and conflicts (as in: "You saw what I was like a couple months ago—I didn't want to be married anymore"). By turns, they encouraged and cut one another off, hoping to build on what previous speakers had said. The energy and goodwill of this group was partly a fluke, I'm sure, but also a by-product of living in the suburbs: these women described more social isolation than their peers who lived in denser areas, and they seemed more grateful to have a regularly scheduled social outlet.

This particular ECFE class included a lawyer, a police dispatcher, a women's basketball coach, a computer scientist, and a Kohl's part-time employee. Just over half of the women had temporarily given up their jobs to care full-time for their infants and toddlers; the others worked part-time, trying to balance work and home, which in almost every retelling was like trying to stand on top of a bowling ball.

At twenty-nine, Angie, whom you met in the introduction, was one of the youngest women in the group. She was also one of the few whose husband, Clint, sometimes attended the class, though it met during the day. "Can I go first?" she asked. "These last two weeks have been the worst two weeks of my life. "Eli"—short for Elijah, her older child, three years old—"had the stomach flu, and he hasn't been sleeping, and I've had the brunt of *everything.* I'm the one getting up with the kids, getting them ready, still working, not sleeping, housecleaning." Her voice quavered a little. "Me and my husband, the relationship is just *horrible* now. He doesn't understand I'm at my breaking point. Yesterday he had this little stomachache, but I had to do everything still. And I was like, *really?*" Her voice broke. "I mean, okay, you have a stomachache. But who cares?"

She started to cry. "And I'm a nurse!"

It was a deliberate punch line, designed to alleviate her self-consciousness, and it worked. Several women burst out laughing. She joined them and briskly wiped away her tears. "He thinks that just because he works five days a week, from five in the morning until two, and because he takes out the garbage—"

"He takes out the garbage?" interrupted one of the women. "Awesome!"

"—or because he does the snow removal or takes care of the water softener," she continued, "that I should take care of the kids more than he does."

"And does he say, 'Oh, it's because they're begging for Mommy anyway?' " asked another woman. "Because my husband says, 'They won't let me help,' and I'm like, *If you'll take the time to do it. . . .*"

"My husband has the 'I make the money, you should do everything else' complex," said yet another. "He's like, 'I've worked all day,' and I'm like, *Gee, I wonder what I've done.*"

"Just, the resentment builds up," said Angie. "And then I'll talk to him about it, and he's like, 'Well, you need to do this and this and this, and then maybe I'll feel better, and I'll take more responsibility for the kids.'"

"Does he know it's not a barter system?" asked a fourth.

"We go back and forth on what we both need," Angie explained. "And then it's okay for a few days. But then we're right back where we started."

"You," declared yet another woman, "have to have a 'Come to Jesus' with him."

With that, the matter was settled. The judgment was definitive, coming from her. She was the police dispatcher.

NEXT TO THE ABRUPT modification of our personal habits, perhaps the most dramatic consequence of having children is the way they alter our marriages. It can hardly be an accident that the first famous paper to challenge the conventional wisdom about the psychological benefits of having children, E. E. LeMasters's 1957 "Parenthood as Crisis," looked at couples rather than mothers or fathers individually.

LeMasters found that 83 percent of all new mothers and fathers were in "severe" crisis. If this figure sounds excessive, that's because it probably is: no one since has posited anything quite so dire. But scanning the contemporary research on the transition to parenthood is nevertheless a pretty sobering exercise. In 2009, four researchers analyzed a variety of data from 132 couples and found that roughly 90 percent of them experienced a decline in marital satisfaction after the birth of their first child—though the change, to be fair, had mainly "a small to medium negative effect" on their functioning. In 2003, three researchers reviewed nearly 100 surveys examining the correlation between children and marital satisfaction and found that "only 38% of women with infants [had] higher than average marital satisfaction, compared with 62% of childless women." In *When Partners Become Parents,* published in 1992, the pioneering husband-and-wife team of Carolyn and Philip Cowan reported that nearly one-quarter of the 100 or so couples in their longitudinal survey indicated that their marriage was "in some distress" when their child hit the eighteen-month mark. "Couples in our study who felt upbeat," they wrote, "were decidedly a minority."

The Institute for American Values points out that one is more likely to be happy raising children as part of a couple than raising them alone, and that's true. It's also true that most marriages tend to decline over time, children or not. But pretty much all research shows that the marital satisfaction curve bends conspicuously the moment a child is born. Some studies say that parenthood merely hastens a decline already in progress, while others say that parenthood exaggerates it. Still others suggest that levels of marital satisfaction are a function of how old the couple's children are, with the early years being an especially challenging time, followed by a period of some relief during the elementary school years, followed by another plunge during the slings and arrows of adolescence.

Yet there's surprisingly little discussion about *any* of these theories in mainstream parenting books, other than cloying bromides (schedule date nights!). In social science, on the other hand, a couple's transition to parenthood is one of the rare subjects that elicits intimate details from researchers themselves. Virtually everywhere else, *When Partners Become Parents* is an academic work, a book-length exposition of years of rigorous interviews and data collection. But its opening pages are intensely personal. The Cowans describe meeting as teenagers, marrying young, and having three children in quick succession. "By the time our children were in elementary school," they wrote, "there was no avoiding the issue: Our relationship was very strained." A number of friends, they noticed, were struggling too:

> As we listened to the pain and disenchantment that other husbands and wives described in their relationships and struggled to make sense of our own, we began to hear a common refrain. We were experiencing distress now in our relationships as couples, but almost all of us could trace the beginning of our difficulties back to those early years of becoming a family.

Before they become parents, the partners in a couple often think of children as matrimonial enhancers, imagining that introducing them into their relationship will strengthen it and give it reasons to endure. And couples with children are, in point of fact, much less apt to divorce, at least while their children are young. But they're also much more prone to conflict. The Cowans note in their book that 92 percent of their sample couples reported more disagreements after their baby was born. (This pattern isn't confined to heterosexual relationships either: a 2006 paper reported that lesbian couples also showed increases in conflict once their children were born.) In 2009, an elegantly designed study by a trio of psychology professors showed that children generate more arguments than any other subject—more than money, more than work, more than in-laws, more than annoying personal habits, communication styles, leisure activities, commitment issues, bothersome friends, sex. In another study, the same researchers found that parents also argue more intensely in front of their children, with fathers showing more hostility, mothers showing more sadness, and the fights themselves resolving with less grace.

E. Mark Cummings, one of the authors, suspects that the reason for such open conflict is fairly straightforward: "When parents are *really* angry, they don't have the self-control to go behind closed doors." And maybe it's as simple as that. But I have another theory, one that's born less of quantitative analysis than of personal experience and interviews with strangers: I suspect that parents argue more aggressively in front of their children because children are a mute, ever-present reminder of life's stakes. A fight about a husband's lack of professional initiative or a wife's harsh tone with her daughter is no longer just a fight about work habits or disciplinary styles. It's a fight about the future—about what kind of role models they are, about what kind of people they aspire to be, about who and what they want their children to become. *Do you want your son to see a father who finds the world an intimidating place and doesn't have the gumption to ask for a raise? If your daughter turns into a screamer when she grows up, from whom do you think she'll have learned it?*

Whatever the explanation, we know there are many potential reasons for relationship conflict after the birth of a child. Increased financial tensions. A totally realigned social and sex life. The sense that the couple is struggling in this thing—*this huge thing*—alone. This chapter looks at all of these issues, but the one I'd like to start with is seemingly banal, yet nearly universal: the division of household labor. When a child comes along, the workload at home explodes exponentially, and the rules regarding who does what and how often get thrown into tense disarray. These rules are much harder to sort through than most couples realize, in part because we lack blueprints for what to do now that both men *and* women are breadwinners. But they're also harder to sort through because they stir up feelings that are about so much more than chores.

women's work

The morning I show up at Angie's home in Rosemount, Minnesota, she too is exhausted, just as Jessie was, but not because she worked the evening before. Angie spent the night alternately struggling with an ailing back and a crying one-year-old, and she had little luck soothing either one. The one-year-old in question, Xavier ("Zay"), is in her arms as she opens the door ("he'd cry if I put him down"), and Eli, her three-year-old, is eating dinosaur oatmeal on the back deck. We walk outside to join him. He's a serious young man, thoughtful and focused and awfully spiffy in his new crew cut. Angie rubs his head and tells him to hurry up. A few minutes later, the four of us pile into the car and head to Little Explorers, a local summer program that meets twice a week.

As was the case with Jessie, I haven't seen Angie since her ECFE class a few months ago. And like Jessie, Angie talks about the challenges of her life candidly and without self-pity. But that's not the reason I'm here. I'm here because Angie and her husband, Clint, both do shift work, and shift work considerably aggravates the challenges of keeping a marriage intact while raising small children. It makes each parent feel like a single parent, with each tending separately to the kids and then heading off to a job without any help from the other. The arrangement is a formula for exhaustion, and it creates a scarcity economy on days off, pitting spouses against one another over who gets the easier assignments on the to-do list and who gets the spare hour for a bike ride or a nap. Each parent is convinced that he or she has had the more difficult week. "We're in the same family with two different lives, two different views, two different opinions," Angie tells me. "What I think is the situation and the hard parts, he doesn't always." And vice versa.

What's interesting is that many couples with young children say they're leading separate lives, even if their schedules are synchronized: they each take different children in the mornings and evenings; on weekends, they split carpooling and chores. The difference is that it's structurally predetermined in Angie and Clint's case. In their situation, many couples can see the same crude outlines of their own, but magnified to the power of ten. "Right now our life is such fragile chaos," says Angie. "If there's something a bit over the norm—Zay not sleeping at night, the dog getting sick, my back going out—it throws everything out of whack."

At the time of my visit, Angie was working every other night as a psychiatric nurse, leaving home at 2:30 in the afternoon and returning home around midnight. Clint, meanwhile, worked five days a week as the morning manager of the Avis and Budget locations at the Minneapolis/St. Paul airport. Every day, he rose at 4:00 A.M. and every day, he returned home at roughly 2:15 P.M. Several times a week—like today—Clint and Angie cross paths for only fifteen minutes.

Angie and I drive Eli to camp. As we're getting into the car, I ask Angie how things have been since her ECFE class a few months earlier, when she seemed so distraught. "Last night, Clint and I got in a bit of an argument, actually," she answers. Her blue eyes are surprisingly alert for a woman who slept only two hours the previous night. "I asked him for help with the dog," she explains, "and he was like, 'Don't look at me!' "

The puppy, Echo, was her idea. She thought the kids would love to have a dog, and she was right. The trouble is, Clint thought house-training a new dog was crazy at this stage in their lives, and he was right too. "So I said, 'Well, then *you* can get up at night with the kids,' " says Angie. And he did, for a while. "But then the baby had a screaming fit at 3:00 A.M.," says Angie, "and that was me."

Why not Clint?

"I felt bad!" she says. "He didn't wake up, and the baby was screaming. But then I had a back spasm, and *I* started crying. . . ." So Clint climbed out of bed and got Angie an ice pack. "Tonight," she declares, "he's going to get up with that kid the entire time."

All this haggling, of course, makes one wonder: did she and Clint discuss how they were going to divide up their responsibilities before the children were born?

"Yeah!" she exclaims, without hesitation. "And he was like, 'Fifty-

fifty! I want to do everything!'" I hear no bitterness in her voice. Just frustration. "It's just that he's still very selfish with his time. Whereas I'm like, 'Kids first.'"

WHEN ARLIE RUSSELL HOCHSCHILD'S *The Second Shift* was published in 1989, it made a startling argument: if one combined their paid and unpaid labor, employed women of the 1960s and '70s worked a full month extra—*of twenty-four-hour days*—over the course of a year. That's not true today. Women are doing far less housework than they used to, and men are doing more; fathers also do more child care; and mothers put more hours into the workforce, in greater numbers. (In 2010, 50 percent of mothers of three- to five-year-olds worked full-time.) As Hochschild noted in an updated introduction to her book—and as Hanna Rosin's 2012 *The End of Men* made persuasively clear—men's economic fortunes have fallen relative to those of women during the last few decades, based on declines in manufacturing jobs. Ideas about who ought to do what in the marital economy have also evolved. In 2000, nearly one-third of all married women reported that their husbands did more than half the housework, versus 22 percent in 1980. Over the same twenty-year span, the number of husbands who did no housework at all dropped by nearly half.

In fact, according to the American Time Use Survey—the gold standard in time measurement—men and women today work roughly the same number of hours per week, though men work more paid hours and women more unpaid hours. This updated calculation led *Time* magazine to wonder, in a 2011 cover story called "Chore Wars," if women were protesting too much about their load.

The reason Hochschild's book became famous, however, probably had little to do with a mathematical equation. Above all else, her book was a series of novelistic portraits of marriages and the tensions embedded in them, as each couple struggled to find a new equilibrium in a culture that offered few guides. Certainly there were examples of egregiously lopsided labor divisions (like that of Nancy Holt, who consoled herself with the declaration, "I do the upstairs, Evan does the downstairs," when "the downstairs" meant the garage, the car, and the dog while "the upstairs" meant everything else). But what made *The Second Shift* so powerful was its analysis of the myths and delusions that couples needed to keep their marriages together. Hochschild could see that repeated—and often touchy, and sometimes failed—attempts to recalibrate the workload had terribly messy emotional consequences. "When couples struggle," she writes, "it is seldom simply over who does what. Far more often, it is over the giving and receiving of gratitude." Toward the end of the book, she elaborates:

> The deeper problem such women face is that they cannot afford the luxury of unambivalent love for their husbands. Like Nancy Holt, many women carry into their marriage the distasteful and unwieldy burden of resenting their husbands. Like some hazardous waste produced by a harmful system, this powerful resentment is hard to dispose of.

And this resentment still persists in marriages to this day, albeit in subtler and different forms. The Cowans, who have been studying the effects of children on marriage for over thirty-five years, say their research shows that the division of family labor is the largest source of postpartum conflict. In *Alone Together*, a 2007 compendium of all sorts of intriguing marriage data, Paul Amato and his colleagues cite a study that shows "household division of labor being a key source of contention between spouses." (Mothers of children ages zero to four, they add, report the most acute feelings of unfairness.)

But perhaps the most intriguing tidbit about domestic fairness comes from a massive UCLA project in which researchers spent more than a week inside the homes of thirty-two middle-class, dual-earner families, collecting 1,540 hours of video footage. The result was a mother lode of data about families and their habits, and it generated dozens of studies. In one of them, the researchers took saliva samples from almost all of the participating parents, hoping to measure their levels of cortisol, the stress hormone. The researchers found that the more time fathers spent in leisure activities while they were at home, the greater their drop in cortisol at the end of the day, which came as no surprise; what did come as a surprise was that this effect wasn't nearly as pronounced in mothers.

So what, you might ask, *did* have a pronounced effect in mothers? Simple: Seeing their husbands do work around the house.

OUR CONTEMPORARY DIVISION OF labor may be getting more equal overall, but it's still unequal for plenty of mothers. As the *Time* story noted, mothers of children under six still work five more hours per week than fathers of children under six. That's not a small difference. In many cases, that time is devoted to nocturnal caregiving, which, as we saw in chapter 1, can be devastating to the body and mind. In 2011, Sarah A. Burgard, a sociologist at the University of Michigan School of Public Health, analyzed data collected from tens of thousands of parents. In dual-earner couples, she found, women were three times more likely than men to report interrupted sleep if they had a child at home under the age of one, and stay-at-home mothers were *six* times as likely to get up with their children as stay-at-home dads.

Funny: I once sat on a panel with Adam Mansbach, author of *Go the F**k to Sleep.* About halfway through the discussion, he freely conceded that it was his partner who put his child to bed most nights. That said so much, this casual admission: he may have written a best-selling book about the tyranny of toddlers at bedtime, but in his house it was mainly Mom's problem.

But let's say, for argument's sake, that a husband and wife *do* work the same number of hours. That is not, in and of itself, an indicator of fairness. In the context of marriage, fairness is not just about absolute equality. It's about the *perception* of equality. "Parents' satisfaction with the division of the child-care tasks," note the Cowans in *When Partners Become Parents,* "was even more highly correlated with their own and their spouses' well-being than was the fathers' *actual* amount of involvement [emphasis mine]." What a couple deems a fair compromise in any situation is not necessarily how an outsider would adjudicate it. They determine fairness based on a combination of what they need, what they think is reasonable, and what they think is possible.

But that's also where things can get awfully complicated. Men and women may, on average, work roughly the same number of hours each day, once all kinds of labor are taken into account. But women, on average, still devote nearly twice as much time to "family care"—housework, child care, shopping, chauffeuring—as men. So during the weekends, say, when both mothers and fathers are home together, it doesn't look to the mothers like their husbands are evenly sharing the load. It looks like their husbands are doing a lot less. (Indeed, in another analysis of those 1,540 hours of video data, researchers found that a father in a room by himself was the "person-space configuration observed most frequently.")

There are some women who'll cheerfully say that if their partners are working more paid hours during the week, they've earned their extra rest over the weekend. But for many mothers, it's not that simple. Paid work, both literally and figuratively, is generally assigned a higher value by the world at large, which has all sorts of unquantifiable psychological rewards. Perhaps just as important, not all work is created equal: an hour spent on one kind of task is not necessarily the equivalent of an hour spent on another task.

Take child care. It creates far more stress in women than housework. (As one woman in an ECFE group put it: "The dishes don't talk back to you.") About two-thirds of the way through *Alone Together*, the researchers actually quantify this distinction, noting that if a married mother believes that child care is unfairly divided in the house, this injustice is more likely to affect her marital happiness than a perceived imbalance in, say, vacuuming, *by a full standard deviation.* Data also make clear that a larger proportion of a mother's child care burden is consumed with "routine" tasks (toothbrushing, feeding) than is a father's, who is more apt to get involved in "interactive" activities, like games of catch. There are differences in the *kinds* of child care that parents do, in short, even if it's all labeled "child care" by researchers attempting to quantify it. (Ask any parent which type of child care they prefer.)

It is, of course, the nature of practically all humans to overestimate how much work they do in any given situation, rather than underestimate it. But when it comes to child care, the women's estimates do seem more accurate. In *Alone Together,* the authors note that fathers guessed that they did, on average, about 42 percent of the child care in their families, based on a large national survey conducted in 2000. Mothers, by contrast, put their husbands' efforts at 32 percent. The actual number that year was 35 percent, and it remains roughly the same today.

These distinctions may explain why women remain so vexed about the family economy, even if they're no longer being shortchanged in absolute terms. "I'm pretty sure Clint *thinks* he does 50 percent of the work when we're home together," says Angie as we're driving along in the car, "but not necessarily the child care. And that's what makes me most stressed."

deadlines and divided time

It's later in the morning. Eli is still at Little Explorers, and Angie is folding laundry on the landing at the top of the stairs. Zay starts to fuss in his crib. Angie pops up to check on him, then returns. "I never get time to put laundry away," she says. "I try. But usually we're just moving it from the clean basket to the dirty basket." Zay is crying now. "Yes, yes, *yessssss,* I hear you!" She jumps up and goes into his room. "Shhhhhhhhh."

Her efforts to mollify him aren't successful. A few moments later, she brings him out and puts him next to her. She resumes folding for a third time, integrating her efforts with peek-a-boo games, just as Jessie had. She tosses a blanket over her son's head. "Where's Zay?" Fold. Toss. "Where's Zay?" Fold. Toss. "Where's Zay? . . ."

This is another thing that quantitative studies of American time use cannot show you: for the majority of mothers, time is fractured and subdivided, as if streaming through a prism; for the majority of fathers, it moves in an unbent line. When fathers attend to personal matters, they attend to personal matters, and when they do child care, they do child care. But mothers more often attend to personal matters while not only caring for their children but possibly fielding a call from their boss. In 2000, just 42 percent of married fathers reported multitasking "most of the time"; for married mothers, that number was 67 percent. In 2011, two sociologists provided an even more granular analysis. They found that mothers, on average, spend ten extra hours per week multitasking than fathers, "and that these additional hours are mainly related to time spent on housework and childcare." (When fathers spend time at home, on the other hand, they *reduce* their odds of multitasking by over 30 percent.) The upshot, the authors write, is that "multitasking likely takes a heavier toll on mothers' well-being than on fathers' well-being."

Being compelled to divide and subdivide your time doesn't just compromise your productivity (as we saw in the last chapter) and lead to garden-variety discombobulation. It also creates a feeling of urgency—a sense that no matter how tranquil the moment, no matter how unpressured the circumstances, there's always a pot somewhere that's about to boil over. As it is, most mothers assume a disproportionate number of deadline-oriented, time-pressured domestic tasks (*Prepare breakfast, pack lunch, drop kids off at school; pick them up, take them to piano, take them to soccer, get dinner on the table by 6:00*). In 2006, the sociologists Marybeth Mattingly and Liana Sayer published a paper noting that women are more likely than men to feel "always rushed," and that married mothers are 2.2 times more likely to feel "sometimes or always rushed" than single women without children. (Free time does nothing to ease mothers' feelings of enervation either—it in fact makes things worse.) Fathers, meanwhile, feel no more rushed than men without children. Here's Kenya again, from ECFE:

> I feel a *huge* pressure around five o'clock. I've got to finish what I didn't do. I've got to plan dinner. I've got to keep my daughter happy, I've got to put her to bed. . . . I thought, without working, I'd be like, *Oh, I'll have all this time*. But I feel all this pressure around five. Whereas when my husband comes home, there's nothing he *has* to do.

But perhaps the hardest and most elusive quantity for a time-use survey to measure is the psychic energy that mothers pour into parenting—the internal soundtrack of anxieties that hums in their heads all day long, whether they're with their children or not. That's one of Mattingly and Sayer's more subtle hypotheses: perhaps mothers feel rushed because the sensitive and logistically intensive parts of raising kids—making child care arrangements, scheduling doctor's visits, dealing with teachers, organizing family leisure hours, coordinating play dates and summer plans—fall disproportionately to them. Angie certainly says as much. "When I'm at work," she tells me, "I'm still only 50 percent nurse, probably. You know? Even if I'm dressing a wound or whatever it may be, I'm always thinking, '*Is Clint going to remember to put sunscreen on 'em?*'"

What happens when she's out alone with Clint?

"It's still the kids on my brain," she says. "Even our date nights, when I'm supposed to be 100 percent wife."

It's interesting that Angie attempts to quantify this feeling in a ratio. Some years ago, when Carolyn Cowan was driving home from a meeting with a group of parents, it occurred to her that she ought to ask them to devise a pie chart of their identities. What percentage of themselves did they see as a spouse, as a parent, as a worker, as a person of faith, as a hobbyist?

Women, on average, assigned a significantly larger proportion of their self-image to their mother identity than the men did to their father identity. Even women who worked full-time considered themselves more mother than worker by about 50 percent. This finding didn't surprise Cowan and her husband—nor were they surprised, years later, when they came across a similiar study showing that mothers who carry the child in lesbian couples give over more mental real estate to their maternal identity than their partners.

What *did* surprise the Cowans, however, was what this visualization exercise portended for the hundred or so couples in their sample: the greater the disparity between how a mother and father sliced up the pie when their child was six months old, the more dissatisfied they were in their marriage one year later.

This finding suggests an even larger context to all these fights about the distribution of family labor. How much does each member of the couple psychologically inhabit his or her parenting role? If each parent prioritizes this role differently, their arguments take on a whole new dimension: *How could you not care about this as much as I do? What kind of parent are you anyway? Doesn't family and family time matter to you? Does this not mean the same to you as it does to me?*

social isolation

It's worth noting that children would almost certainly be easier on marriages if couples didn't rely so much on one another for social support. But unfortunately, they do. What this means, all too often, is that parents can feel awfully alone, especially moms.

In 2009, a specialty consulting firm surveyed over 1,300 mothers and found that 80 percent of them believed they didn't have enough friends and 58 percent of them felt lonely (with mothers of children under five reporting the most loneliness of all). In 1997, the *American Sociological Review* published a paper showing that women's social networks—and the frequency of their contact with the people in those networks—shrink in the early years of child-rearing, with the nadir occurring when their youngest child is three. (The expansion thereafter, the authors say, likely has something to do with the new connections mothers make once their children reach school age.) And the most popular form of Meetup in the country, by a substantial margin, is mothers' groups. "That really surprised me," Kathryn Fink, the company's community development specialist, told me in a phone conversation. "Before I worked at Meetup, I assumed that if you chose to be a stay-at-home mom, you could rely on your preexisting social network."

Fink isn't the only one who finds it surprising that new mothers pine for connection. So do many new mothers themselves. The conventional wisdom about children is that they bring together not just couples but extended families, social networks, entire *communities.* There's some evidence suggesting that this is true—eventually. Sociologists who have examined the complex circuitry of American social life have noticed that people with children know their neighbors better than those without children do; they also participate in more civic organizations and form new ties through their children's activities and friends. But these are not necessarily their most intimate or emotionally sustaining ties. In his landmark book *Bowling Alone: The Collapse and Revival of American Community,* Harvard political scientist Robert Putnam explains this distinction by noting the difference between "machers" and "schmoozers": machers are community muckity-mucks, people who make things happen through formal involvements in civic organizations; schmoozers are social butterflies and informal hanger-outers, people with active social lives whose "engagement is less organized and purposeful." If you're young and unmarried and renting, odds are that you're a schmoozer. Once you marry and buy a home, you may continue some elements of your schmoozing life, but you're also more likely to reapportion some of those energies to macherdom.

And children seal the deal. Once women and men become mothers and fathers, their purposeful socializing—through churches or synagogues or mosques, through PTAs, through neighborhood watch groups—goes up, up, up. But informal socializing with friends goes down, according to Putnam. So does socializing associated with leisure interests. "Holding other demographic features constant," he writes, "marriage and children are *negatively* correlated with membership in sports, political, and cultural groups [emphasis his]."

In the early days of infancy, motherhood can be especially isolating, with mother and child forming a closed loop. Modern social scientists aren't the only ones who have noticed. Dr. Benjamin Spock talked about it over half a century ago. "Women who have worked for a number of years and loved not only the job but the companionship," he wrote in *Problems of Parents,* "often find children quite limited company." He added: "The woman who chafes at the monotony of child rearing (and I'm assuming that most mothers do at times) is really beset from two directions: the separation from adult companions, and being bottled up with the continual demands of the children. I don't think Nature ever intended the association to be quite so exclusive."

The subject of isolation came up a lot in ECFE classes, especially from mothers of newborns and toddlers who had dropped out of the workforce. The women in Angie's class discussed it at length:

SARA: I didn't think I'd feel as *alone* as I have at times. I feel like it's just me and the boys.

KRISTIN: Me neither. My mom probably gets annoyed, because I call her more than I should. I feel like that's my connection.

ANGELA: Yeah, and I sort of thought, *Well, I'm not around people most of the day anyway, I'm stuck in a cube. How can it be that different?* But it is different, because when I was at work, I could stand up and talk to *adults.*

The real surprise to me, however, was the testimony of stay-at-home fathers. Almost to a man, the stay-at-home dads I met in Minnesota described how challenging it was to find a network of compatriots in their brave new roles. "The first year, I was incredibly isolated," a father told his group in a fairly representative moment. "I felt weird about hanging out with other moms. I didn't feel like I could approach them in the same way. I mean, if my wife were staying at home, she could have. But me . . ."

So what did he do?

"I was really, reaaaaaally nice to other dads I met at the park."

THERE'S A LARGER CULTURAL backdrop to this loneliness. Today's parents are starting families at a time when their social networks in the real world appear to be shrinking and their community ties, stretching thin. Yes, mothers and fathers may have many friends on Facebook, and Facebook is an invaluable resource for them in all sorts of ways, whether it's crowd-sourcing questions about relieving colic or simply posting a comment that helps unspool a thread of sympathy (like Angie's post of October 2011: "I should be sleeping").

But our non-virtual ties are another matter. In 2006, a survey in the *American Sociological Review* famously reported that the average number of people with whom Americans could "discuss important matters" dropped from three to two between 1985 and 2004, and that the number of Americans who felt they had no confidants at all had more than doubled, from 10 to 24.6 percent. The far better-known chronicle of American solitude, though, is *Bowling Alone,* in which Putnam manages to document the decline of almost every measurable form of civic participation in the waning decades of the last century. When the book came out in 2000, critics complained that Putnam had focused too much on obsolescent activities (card-playing, Elks club meetings) and given short shrift to new forms of social capital, like Internet groups. (Facebook hadn't even been invented back then.) It didn't matter. The book still resonated with politicians and laypeople alike, and if my conversations with parents are any indication, Putnam's findings and themes still resonate deeply with families today, in spite of their vast virtual networks.

Take our dwindling neighborhood ties: during the last quarter of the twentieth century, according to Putnam, the number of times married Americans spent a social evening with their neighbors fell from roughly thirty times per year to twenty, and subsequent studies have shown that this number continued to drop through 2008. "When I first moved to our block," Annette Gagliardi, the veteran ECFE instructor, told one of her classes, "I didn't know anyone, and my mom was several towns away. So the older women on the block pulled me in. They were the ones I called in the middle of the night and said, 'My child has a fever.'" There's no substitute, she said, for that kind of embodied contact with fellow parents. "Yes, I can text someone," Gagliardi said. "Or yes, I can look online at a parenting website. But that's not the same as someone racing over to my house and teaching me how to put a butterfly bandage on my daughter's wound."

Our relative estrangement from our neighbors is partially an outgrowth of a positive development: more women are in the workforce. With more women heading off to the office in the morning, more houses inevitably sit empty during the afternoon. But our diminishing neighborhood ties cannot be explained by social progress alone. It can be explained by sprawl, which pushes our houses farther and farther apart. It can be explained by anxieties about crime—kidnappings in particular—which have all but obliterated the once-standard practice of sending children out into the yard or the street. Putnam, like his colleagues studying time use, also describes a sensation of "pervasive busyness" among Americans today, a sense that we are chronically and forever feeling rushed.

The net result has been the death of the "pop-in," to recruit a term of Jerry Seinfeld's, whereby the Kramers and Elaines of the world show up unexpectedly at your doorstep bearing gossip and harmless, unhurried conversation. In the mid- to late seventies, the average American entertained friends at home fourteen to fifteen times per year, according to *Bowling Alone;* by the late nineties, that number had split nearly in half, to eight.

ANGELA: When I was growing up, my mother was surrounded by people home alone with their kids. Every afternoon someone was going to call or we were going to visit someone. My mom would load us all into the car. I mean, maybe my mom was just a social person, but—

SARA: No, it was the same in my house: every Sunday we'd load in the station wagon and just go visit someone. And now you feel like you're intruding, because everyone's so busy.

Without the pop-in, without the vibrant presence of neighbors, without life in the cul-de-sacs and the streets, the pressure reverts back to the nuclear family—and more specifically, to the *marriage* or partnership—to provide what friends, neighbors, and other families once did: games, diversions, imaginative play. And parents have lost some of the fellowship provided by other adults.

Of course, raising children would be easier on marriages if we still lived in extended-family groupings. But as Stephanie Coontz notes in *The Way We Never Were*, "extended families have never been the norm in America." (The highest percentage of people living in extended families on record was just 20 percent, and that was between 1850 and 1885.) What is true, though, is that college-educated Americans tend to live farther away from their parents than those who have only completed high school. In marriages where both partners finished college, the odds are just 18 percent that they live within thirty miles of both their mothers. (Among the high school–educated, the odds increase to 50 percent.) Education clearly results in mobility, which almost by definition weakens family ties.

Weakened family ties have all sorts of consequences for parents. They affect, for instance, women's workforce attachment: married women with kids in elementary school or younger are 4 to 10 percent more likely to work if they live near their mothers or mothers-in-law. The social lives of parents are also affected: without the most reliable, most psychologically reassuring, and (above all) most affordable form of babysitting—namely, grandparents—a simple evening out with one's spouse is a much harder sell.

"I do have an aunt who lives just fifteen minutes away," Angie tells me, when I ask whether she has a network of caretakers to rely on. But that's it. Everyone else is far away or in poor health. Angie and Clint are part of the so-called sandwich generation, the generation inconveniently squeezed between aging parents and young children, meaning they face caretaking stresses no matter which direction they crane their necks. As Americans live longer and women defer childbearing into their thirties, this generation is only expected to grow.

disobeying orders

It's lunchtime, and Eli is sitting in front of a plate of Angie's chicken parmesan. He is not, however, eating it. He is instead contemplating his polar bear mug.

"What do polar bears eat?" he asks.

"Fish," Angie replies.

"What else?"

"I don't know. Would you eat, please?"

He doesn't. Angie looks at him. "You're not eating again until dinner. If you don't eat, you won't have a snack."

Eli tries picking off a small piece with his fingers.

"Please use your fork. Is that a polar bear bite?"

"I'm eating like Zay," he answers. Zay gets to eat with his fingers.

"If you eat like Zay," says Angie, "that'd be great. Zay's *eating.*"

Eli has an idea. "Watch this, Mama!" He tips his plate toward his mouth and some spaghetti slides in.

"Eli, use your fork."

"I can't."

"Why not?"

"Because I just did it this way."

Angie stands up, shrugs, and moves on to other things. "As long as you get food into you."

ALL PARENTS FIND THEMSELVES in absurdist loops of non-argument with their children. At their most benign, these disputes are merely annoying; at their worst, they're outright maddening. It's not surprising that the parenting section of bookshops is filled with guides to coaxing obedience from children. What is surprising, though, is how little of them cite the behavioral research on this subject. If you dive into it, you'll discover that *all* American parents, even well-adjusted ones, spend a staggering amount of time each day trying to get their toddlers and preschoolers to do the right thing—as often as twenty-four times an hour, according to some studies—and that toddlers and preschoolers, even well-adjusted ones, spend a staggering amount of time resisting these efforts.

It may seem strange to bring up studies about child compliance in a chapter about marriage. But not if you consider one very salient fact: almost all of these efforts to get children to comply are made by mothers, not by fathers, and this asymmetrical dynamic can add a low-frequency hum of resentment to a relationship, because Mom gets the job of family nag. She didn't seek this job either. It's a simple matter of numbers: if mothers spend more time with their children than fathers do, they're bound to issue more commands. (*Put your shoes on. Are you going to pick that up, or are you waiting for the house elf to get it? Where on earth did you find that, and whatever it is, please take it out of your mouth.*) Even more insidiously, compliance requests tend to be about time-sensitive matters. (*Put on your coat, we have to leave. Brush your teeth, it's getting late.*) And mothers feel quite rushed as it is.

The first time I came across data about compliance requests was in a 1980 paper titled "Mothers: The Unacknowledged Victims." The name pretty much says all you need to know. The author's first conclusion was that, during the preschool phase, "rearing normal children provides the mother with high rates of aversive events," which happened as frequently as once every three minutes, according to his review of the literature.

But this study was hardly the only one. There was the 1971 study from Harvard that I mentioned in the introduction, which found mothers correcting or redirecting their toddlers every three minutes, and their toddlers listening only 60 percent of the time. Three years later, researchers from Emory and the University of Georgia found that psychologically healthy kindergartners from higher-income homes listened to their mothers only 55 percent of the time, and children from lower-income homes, 68 percent. (The mothers of lower-income children consistently issued more orders.) And these studies are dotted throughout the social science archives, all the way to the present day. In one of the more recent papers I peeked at—from 2009, this was—mothers and toddlers were averaging a conflict every two and a half minutes.

Naturally, there are limits to how seriously one should take these kinds of studies. In the words of Urie Bronfenbrenner, who helped found Head Start: "Much of contemporary developmental psychology is the science of the strange behavior of children in strange situations with strange adults for the briefest possible periods of time." But they were a delight to discover nonetheless. Who knew that my son's dissident behaviors—and my responses—were so commonplace?

Pamela Druckerman, author of the 2012 best-seller *Bringing Up*

Bébé, would argue that American mothers frequently lock horns with their children because they don't know how to discipline them with the same firmness that the French bring to the task. No doubt there's some truth to this observation. How children behave is always culturally mediated. But what interests me is that *mothers* give most of the orders, and this compliance literature makes it clear that giving those orders is taxing and stressful. In the all-mom ECFE groups, the subject came up all the time. In a class just before Angie's, two women had this exchange:

KATY: I have night classes, so I have a list for my husband before I leave—be sure you give our son a bath, be sure you put him in his jammies. And I'll come home four hours later, and they've both fallen asleep on the floor, all clothed, and there's a movie playing and a bag of chips.

COURTNEY: Same here. I think my husband thinks of parenting as play, and I see it as work.

KATY: Or watching the two of them grocery shopping—that's awful. Whatever my son wants, my husband gets it for him.

The next day, in a different class:

CHRISSY: My husband will give the kids peanut butter and jelly and yogurt and be like, "Woohoo! Dinner!" And I'm racing to put out the vegetables, saying, "Uh, guys, you have to eat these too."

KENYA: I know! I don't know why my husband winds up being the fun guy. I come home, and my daughter tells me, "Daddy lets me have pop."

At that point, the instructor, Todd Kolod, felt compelled to intervene.

"May I just speak up on behalf of dads?"

The women smiled. *Sure.*

"I think they need the chance to make mistakes," he said. "They'll say that they'll try to help with laundry, and then, *once*, they'll ruin some item they were supposed to hand-wash, and they're cut off from doing the laundry forever."

The women agreed he might have a point.

And he did. All relationships benefit from generosity. (Nor do kids stop growing if they're fed peanut butter and jelly for dinner.) But the women had a point too. What they were responding to, really, was Daniel Gilbert's observation from chapter 1. "Everyone is moving at the same speed toward the future. But your children are moving at that same speed with their eyes closed. So you're the ones who've got to steer." Typically, it's mothers who take the wheel.

It's exhausting to be the family compass and conscience. It means the stuff of everyday life becomes a source of tension; it means you're the designated family prig. *I don't know why he's the fun guy.* When Kenya said this, she didn't sound angry. She sounded sad.

Here lies yet another explanation for the happiness gap between mothers and fathers. It's not necessarily the quantity of time mothers spend with their children that's the problem. It's how they spend it.

who's having sex?

As mercilessly unsentimental as it is to say this, children would have less of an impact on marriage if the institution itself weren't so heavily burdened by romantic expectations—which, as we saw in chapter 1, are relatively new. Before the late eighteenth century, marriage was a public institution, inseparable from raising families and binding individuals to the broader community. But sometime around the late eighteenth century, as Jane Austen was completing her first draft of *Pride and Prejudice,* a different idea began to take shape: marriage was for love. Today, 94 percent of singles in their twenties believe that spouses should be soul mates "first and foremost," according to a 2001 Gallup poll, while just 16 percent believe that children are the primary objective of marriage.

This redefined notion of marriage—as a sheltered loop of mutual fulfillment rather than a public institution for the commonweal—generated an inspired term from the sociologists David Popenoe and Barbara Dafoe Whitehead. They called it a "SuperRelationship," which they defined as "an intensely private spiritualized union, combining sexual fidelity, romantic love, emotional intimacy and togetherness."

If most of us begin our marriages with these expectations, is it any wonder that we experience children as a disruption?

Lots of couples genuinely enjoy their coupledom. Unlike the literature about raising children, many studies about marriage suggest that the institution makes people happier and more optimistic (though it's possible that happier people get married in the first place). Studies also suggest that married people are healthier.

So what, precisely, gets compromised when a child enters the picture?

Well, time alone together, famously (hence those endless exhortations to schedule date nights). Estimates vary as to how much a couple's time together declines, but the most commonly cited study says it drops by two-thirds once a child is born. The nature of this time together changes dramatically too. Social scientist and St. Paul couples therapist William Doherty, who is also an adviser to ECFE, likes to tell the story of a beautiful couple, marvelous country-Western dancers both, who came to his office for counseling one day. They'd met as young adults at a dance in Oklahoma; when they were dating, they'd go out dancing all the time, and other couples would inevitably form a circle around them, just to watch them spin. At some point Doherty casually asked them when they'd last gone out dancing. Their answer? Their wedding reception, twelve years earlier.

Almost everyone seems to agree that a couple's sex life also changes after children come along, though it's surprisingly difficult to find strong data supporting this hypothesis. A few studies do manage, however, to confirm the suspicion that this is true, either indirectly or by design. A paper from 1981, for instance, looked at 119 first-time mothers and found that 20 percent were having sex less than once a week at their child's first birthday, while only 6 percent were having sex that infrequently in the three months prior to conception. (Then again, they may have been actively trying to get pregnant during those three months—and hence, having more sex.) Another small study, conducted slightly later, found that a child, along with "jobs, commuting, house-work . . . conspires to *reduce* the degree of sexual interaction" in the early years of marriage, "while almost nothing leads to increasing it."

In 1995, a much larger study concluded that the presence of young children—specifically four years of age or less—has an even more substantial impact on a couple's sexual frequency than pregnancy itself (and only slightly less significant an impact than poor health). Having five- to eighteen-year-olds at home, on the other hand, slightly *increases* sexual frequency. (Though here's a question: if the authors had analyzed *just* the parents of adolescents, would they have come to the same conclusion? Because teenagers can pose a true circadian dilemma, springing alive in the wee hours like so many vampire bats. It makes the prospect of nighttime congress particularly dicey.) And here is my favorite detail from that study: "Respondents with low and high educational attainment levels reported less frequent marital sex. This curvilinear relationship is taken into account in all analyses." Make of that what you will.

But it's very hard—almost impossible—to find concrete numbers about the frequency of marital sex once babies enter the picture. In an evening ECFE class of working fathers, the instructor, Todd Kolod, surprised everyone by asking the question outright: how much sex is it realistic for a dad with young children to expect? Everyone paused for a moment, trying to gauge whether they should answer this question seriously or punt with a joke.

FATHER #1: Whenever you can talk her into it.

FATHER #2: Can we put it this way? What's realistic about going out to the movies? It's like once . . . a year.

TODD: We don't really know what's realistic, do we? That's part of the problem. But really, seriously: what do you think?

FATHER #3: Some of our friends, they started out crazy. One friend, for years, I called him "nine times." And now, I'll talk to the same guy, and he'll be like . . . nothing.

FATHER #2: Okay, let's make it more uncomfortable: how many times do we masturbate a day?

FATHER #4: Ha-hey, speak for yourself. . . .

But numbers may be beside the point. If you really speak to men and women about this, both alone and in groups, they'll tell you that sure, they miss their old erotic selves, those people who once could only be coaxed from bed in order to pee or eat. But in many cases, those selves were fading away even before the baby came along. (There's evidence that the most precipitous drop in sexual frequency occurs just after the "honeymoon" year of marriage—a sobering thought.) What most couples really seem to miss is that sense of closeness and aliveness that sex brings. "I don't think I had ridiculous expectations about intimacy," a father told me. "And maybe it's easier for guys, in a way, because we can look at a woman and say, 'She doesn't *look* exhausted and wiped out. She *looks* like she's back to her old self.'" Whereas his wife's attitude was *I am exhausted. Can't you let me sleep without making me feel guilt for denying you something?* It took him a while to realize this. "Honestly, sex itself wasn't even what was driving it for me," he said. "It was our lack of connection. And the less connected I felt, the more I felt like I was going to snap."

"I think it's about mastering the art of the quickie," Angie tells me. "It's like, 'Okay! The kids are sleeping! And I've got to go to work!'" She makes a chop-chop with her hands, then smiles. "We try for at least once a week, if not more. If it gets any longer than that, we just don't feel . . ." (and then that word again) " . . . *connected.*"

Yet that connection comes at a price, just as ignoring it comes at a price. "In our erotic lives we abandon our children," writes Adam Phillips, the British psychoanalyst, in *Side Effects,* "and in our familial lives we abandon our desire." When faced with this unsavory dilemma, observes Phillips, "most people feel far worse about betraying their children than about betraying their partner."

Another woman from ECFE:

"It's funny: my husband has been asking me for quickies lately—it's been a couple weeks. And I was like, *I cannot give to another person.* And he is the one, unfortunately, that has to make the sacrifice. He's the one I can say no to. But I probably should give in, because it's good for *us.*"

When forced to choose between her husband and her kids, she chose the kids.

But here's some news that ought to reassure this mom and all others who have opted to remain in the workforce—as well as dads, for that matter, who put in very long hours: A 2001 study in the *Journal of Sex Research*, which looked at a sample of 261 women with four-year-olds, concluded that "there were no differences between homemakers and women employed part, full, or high full time for several measures of sexual functioning. Neither were there differences between husbands employed full and high full time." (High full-time, in their estimation, was fifty-plus hours per week.) Rather, the authors found, what played the most powerful role in determining the quality of a couple's sex life was a deceptively simple idea: the importance of the marriage to each partner's identity. The more central each one found it, the more satisfied he or she was. Believing in marriage, at least if you're in one, turns out to be the most powerful aphrodisiac of all.

men's work

It's 2:35 P.M., and Clint, a sweet-faced fellow with a barrel chest and a serious disposition, walks quietly through the door, a carabiner of keys tinkling on his belt. He radiates dependability and patience, a quiet belief in hard work; like Angie, he looks tired—he's been up, remember, since four—but manages to move with the speed and energy of someone who's had a full night's sleep. He's wearing shirtsleeves, a tie, and black pants, which he'll swap out in about ten minutes for a charcoal-colored soccer T-shirt and cargo shorts. Angie has just changed into her scrubs. He scoops up a child in each arm, impassively receives an update about each, and kisses his wife hello and good-bye. They all hug for a moment. Then Angie is out the door, and Clint whisks Xavier into the Bumbo baby seat on the kitchen counter. He pulls out some strawberries from the fridge, begins to cut them, and gives a few to the baby and Eli.

"Can I have a surprise snack?" asks Eli.

"You can have strawberries," says Clint.

"Maybe that can be my surprise snack."

"Then it's not a surprise."

He is warm as he says this, but very focused. Later on, I will look in my notes and see that I have written and underlined in all capital letters: *THIS MAN IS ALL BUSINESS.* Not rejecting or disengaged, it should be said, but certainly possessed of a very different style from his wife. When Angie was making lunch only a couple of hours earlier, she left a cheerful mess as she went, often getting pulled away by the boys at just the moments when she'd intended to tidy up. Whereas Clint is immaculate and methodical as he goes. He's so brisk and efficient about washing dishes that it looks like he never dirtied them in the first place.

He opens the fridge and stares into it. "I'm going to think about getting your dinner soon. What did you have for lunch—?"

"Cheese chicken and spaghetti. It was really good, but I didn't like the chicken."

"Why not?"

"It was kind of spicy. I liked the spaghetti."

Clint closes the fridge and goes to get dog food. The baby is quiet, watching and nibbling on strawberries and cereal puffs. Eli goes downstairs to watch the end of an Elmo potty-training video. Clint starts unloading the dishwasher. When he finishes, he takes Zay out of the Bumbo and joins Eli, who's struggling downstairs with a Lego ice-cream truck. Clint relieves him. "Here, let me help ya . . . you put this thing on the back of it." As he assembles the truck Clint plumps with life, like a diabetic who's finally been handed a sucking candy. "I'm biased about Legos," he says, noticing me noticing him and reading my thoughts. "They're what I played with when I was little." He lines up a few animals on a Lego platform for Zay.

This old-school play goes on for a lovely while. Clint explains that he always coordinates a group activity before dinner, so that the kids don't zone out in front of the television; he describes his preference for toys you can build with over chiming plastic geegaws. Play gives him obvious pleasure. But then he glances over at Zay and looks at his phone. "I'm checking the time," he explains. "Trying to figure out when to plan dinner, before The Meltdown."

Eli points to a bus he made out of an egg carton. "Want to make another one with me, Daddy?"

Clint chuckles and gets up. "How 'bout we make dinner first, okay?" His eyes are already on the kitchen. There's dinner to be made, a meltdown to be averted, a nighttime routine to be rallied through. He's on a schedule now. He's all business again.

WATCHING CLINT GO ABOUT his afternoon and evening routine, it is hard not to notice that he and his wife are a portrait in contrasting parenting philosophies—and that their children, in turn, respond very differently to each of them. Zay, for one, could hardly tolerate being put on the ground by Angie that morning. The second she tried it, he bawled. She could have taken a stand, sure, and left him there to tough it out; Clint would say, albeit gently, that Angie has had a hand in creating this predicament, because she allows herself to be manipulated by Zay's distress. ("Zay's not *expecting* me to pick him up the moment he whimpers," Clint notes.) But to leave Zay to cry would only compound Angie's terrible sense that she's not doing all she can do for him, and she feels bad enough going off to work three or four nights a week—the second the kids catch sight of her in her scrubs, they start to cling. So while she's at home, she doesn't put Zay on the ground or in the Bumbo. Instead, she works one-armed and lopsided, straining her back and making the awkward progress of a contestant in an egg toss.

"What I think are the hurdles, Clint often doesn't," Angie told me, not long before she left. "He thinks I cause some of the worry unnecessarily. I think the worst is when he feels helpless." By "helpless," Angie doesn't mean that Clint himself feels overwhelmed. She means that Clint thinks *she's* overwhelmed and there's nothing he can do to soothe her. "When he thinks I can't do something he thinks is simple," she explains.

But of course Angie can do simple things. What's really happening in the moments when Angie seems overwhelmed is that she's fracturing her time. (*Toss.* "Where's Zay?" *Fold. Toss.* "Where's Zay?" *Fold. . . .*) Whereas Clint, both by habit and temperament, is quite clearly the kind of fellow who optimizes his time and probably was doing so even before the kids were born. Parenthood has simply completed his transformation into an efficiency-seeking Scud missile. He acknowledges as much. "Whereas Angie may view something from the feeling aspect, or the enriching aspect—'The kids have to go to the park! They have to spend time doing something different!'—I'm looking at it more from a time-efficiency point of view," he says.

This time-efficiency point of view can be mistaken for a kind of Vulcan-ness. But that's not what it is. It's really the difference between method acting and more classical techniques for getting into a role. Angie approaches parenting intuitively, from the inside out, while Clint approaches it from the outside in. "She just *knows* what needs to get done," says Clint. "Whereas I stumble across it." He thinks about this. "I mean, it's not like if the baby has a poopy diaper, I say, 'Here, you do it!'" He mimes handing off a soiled baby in revulsion. "It's that, if she sees it happen and I don't see it at the same time, she gets upset." He thinks some more. "She's so in tune with the baby monitor, she'll wake up seconds before they do."

I hear this a lot from parents. One—usually the mother—is more alive to the emotional undercurrents of the household. (In *A Home at the End of the World,* Michael Cunningham writes: "She knows something is up. Her nerves run through the house.") The result is that the more-intuitive parent—in this case Angie—sometimes feels like the other parent is not doing his or her fair share, while the other parent—in this case Clint—feels like the intuitive parent is excessively emotional. When really, what may be going on is that the couple is experiencing time differently, because each person is paying attention to different things. When Angie hears the baby monitor or sees that Zay's diaper needs a change, she jumps. Those are time-sensitive tasks, and she's the first responder in the house. Which makes her feel, to use her word, "overwhelmed."

Clint acknowledges this difference. "The way that I come at it is not in real time, for lack of a better term," he says. "I look at the whole picture. I say, if I'm doing 100 percent of the snow removal, the yard work, the maintenance, the dishes, and the meals, you'll have to pick up more slack on the kids' end." He adds that Angie doesn't always appreciate the less time-sensitive contributions he routinely makes. "Maybe she doesn't care about that stuff as much until the dishwasher *breaks,*" he says (which it recently did). "Then *I'm* the one to figure out how to fix it, because she wants those dishes washed."

Yet he's also very attuned to the strain that comes from the moment-to-moment handling of the kids. "That real-time sense that she's doing more? I probably fail to validate that as much as I should," he concedes. He brings up that time in the spring when I first met Angie, and he and the kids were all under the weather. He knew she was exhausted. He knew the kids were sick. He regrets not having jumped in to help. "It was, *Right here, right now, the kids are sick, I want a break,*" he says. He gets it.

At the beginning of each interview with Angie and Clint, I asked them to give a rough breakdown of their household division of labor. And their estimations, for the most part, were remarkably similar: Clint does almost all the cooking. Angie does almost all the night duty, because Clint rises for work at 4:00 A.M. Clint does a little more cleaning; Angie a little more laundry. Angie gets a bit more of the groceries and does the lion's share of shopping for the kids, the doctor's appointments, the extracurricular activities. Clint does the outdoor and household maintenance and 100 percent of the bookkeeping. They each made these same assessments, independently of one another.

The only area over which they disagreed was the one that mattered to Angie the most: child care. She estimated that she does 70 percent, and not because she spends more time at home. She said she did more child care *even when Clint was around.* "If we're just having a home day," she told me, "I do more of the diapers. If Eli's outside, I'm checking and making sure he's okay. I'm keeping the TV off, I'm engaging." Most important, Clint always manages to claim free chunks of time for himself that she never manages to locate for herself. "He can spend two or three hours in front of the computer on the weekend, doing his hobbies," she said. "But recently, I wanted to try this ninety-day boot camp workout, and I couldn't find the hour each day to do it."

Clint answered differently. He said they divide the child care fifty-fifty. "It's push-pull," he said. "If she has a bad day, I do more of it, and if she has to work three shifts in a row, I have to do more of it."

Fifty-fifty and seventy-thirty is a big difference—especially given how little daylight there is between Angie and Clint over everything else. Why, given how sensitive and attuned they are to one another, should this be?

BEFORE PROCEEDING ANY FURTHER, I should pause here to note that this conversation Clint and Angie are having about who does what—this conversation that *all couples* have about who does what—happens at the expense of a more important conversation: does the state have an obligation or moral imperative to help out mothers and fathers? In America we wind up having these arguments privately because our politics allows little room for us to have them publicly. One hates to invoke Sweden at this moment—it really is the most predictable cliché—but some of the happiest parents on the globe are, in fact, in Scandinavia and the other northern European countries with large social safety nets.

In 2012, the sociologist Robin Simon and two of her colleagues measured the difference in happiness levels between parents and non-parents in twenty-two industrialized nations. The country with the greatest gap, by far, was the United States. As a rule, in fact, this difference tended to be larger in countries with less generous welfare benefits and smaller—or inverted entirely—in countries that offer the most support to families.

Arnstein Aassve, a demography professor in Milan, detected a similar pattern in 2013. After examining parental well-being levels across twenty-eight European nations, he and his colleagues concluded that "in general, the happiness that people derive from parenthood is positively associated with availability of childcare." This was especially true in places where child care is available for children between the ages of one and three (France, the Netherlands, Belgium, Scandinavia). In those parts of the world, mothers are consistently happier than nonmothers.

The relationship between access to child care and parental well-being is sometimes deceptive. We cannot necessarily assume that one is the *cause* of the other. Countries with more generous welfare benefits tend to score well on all sorts of social indices: their corruption levels are lower, their gender-parity levels are higher, they tend to offer affordable health care and higher education. To the extent that parents' psychological strains are financial—and many of them are—countries that provide these amenities go a long way toward relieving stress on couples and single parents alike. "These countries," Aassve tells me, "are scoring on a whole range of categories that make people feel optimistic or safe about raising children."

In the opening pages of her 2005 book *Perfect Madness,* Judith Warner writes about what it was like to receive such improbable benefits when she lived in Paris during her early child-rearing years:

> My elder daughter, from the time she was eighteen months of age, attended excellent part-time preschools where she painted and played with modeling clay and ate cookies and napped for about $150 per month—the top end of the fee scale. She could have started public school at age three, and could have opted to stay until 5 P.M. daily. My friends who were covered by the French social security system (which I did not pay into) had even greater benefits: at least four months of paid maternity leave, the right to stop working for up to three years and have jobs held for them.

Meanwhile, a report from Child Care Aware of America notes that in 2011 it cost more for families to put two children in day care than it did for them to pay their rent—in all fifty states.

It's worth imagining how different Angie and Clint's lives might be if they were assured access to the same affordable child care arrangements, and if they both knew they could leave their jobs for a year or three without fear of losing their place in the workforce. At the moment, such luxuries are unthinkable to Americans.

Yet they appear to confer true psychological benefits. In a 2010 study, Nobel Prize winner Daniel Kahneman and four of his colleagues compared the moment-to-moment well-being of women in Columbus, Ohio, to that of women in Rennes, a small city in France. Although the researchers found many similarities between their two samples, the French and American women differed in one very significant way: the French enjoyed caring for their children a good deal more, *and* they spent a good deal less time doing it. In his 2011 book *Thinking, Fast and Slow*, Kahneman speculates that this may be the case because French women have greater access to child care and "spend less of the afternoon driving children to various activities."

"me time"

Clint is in the kitchen. His mission: dinner. He puts Zay in the Bumbo, and Eli climbs onto the counter next to his brother. "What do you want to eat?" Clint asks. "I can grill some chicken, or we can have some shrimp. . . ." He pulls out a box from the freezer and shows it to Eli.

"I just want to make toast."

"Toast is a breakfast food. It's not for dinner."

"I don't like anything."

"This is why you need a nap," Clint tells Eli, picking him up. Eli holds his father's face and, for the first time ever, notices his stubble. "What's that?"

"Hair. I forgot to shave this morning."

"Why are you wearing it?"

"It just grows. On boys. Right there." He points to his chin. "Are you avoiding my dinner question? Are you going to eat my dinner? If you do, maybe you can watch one more show." Eli seems content with this plan. In the meantime, Clint sends him downstairs to clean up his toys.

I ask if this is his usual routine—kitchen prep work first, then some playtime with the boys, then dinner. It has a comfortable rhythm to it, and it's awfully efficient.

"Pretty much," he says. He puts away a last glass from the dishwasher. "That way," he adds, with a mild smile, "later on, I can get in some me time."

ME TIME. SUCH A simple phrase, and yet it reveals a universe of difference between Angie and Clint, and possibly between most mothers and fathers. The majority of parents feel like they don't have enough time for themselves, but mothers are especially burdened by this feeling.

One can easily see this pattern with Angie and Clint. Clint gets home after a long day, and his goal, quite reasonably, is to get the kids situated and to map out the rest of the evening, in the hope of creating a modest stretch of leisure later on. If that means doing mundane housework while the kids are still awake, rather than in bed, so be it. "Whenever the kids are doing something where I don't have to interact with them," he says, "I use the time to do daily chores."

Something where I don't have to interact with them is another telltale phrase. It's not the sort of phrase that would spring to mind for most middle-class mothers, particularly if they're in the workforce. Painfully aware of the time they're *not* spending with their kids, working mothers are more likely to say that they should *always* be interacting with their children once their heels are off. If they're not working . . . well, why are they at home, if not to interact with their children?

Yet Clint is perfectly comfortable leaving his children to their own devices every now and then, and no rational jury of his peers would declare him unloving on account of it. They would simply say he's protective of his time.

Angie, however, has no such attitude toward *her* time. Earlier in the morning, as she put Zay down for a nap, I asked if she, too, wanted to try to nap. She swatted the idea away with her hand. "It's not worth it. I'll only get an hour, and I need like twenty, and there's so much to be done here. . . ."

The Cowans have a word for this feeling. They call it "unentitlement." I thought of it a lot as I watched and listened to Angie. Clint must notice it too. As he's making dinner, I ask him why he thinks he has more free time than Angie does. "Maybe for the same reason she buys more of the kid stuff," he answers, after mulling it over. "When she has money, she feels guilty if it goes to her, but if it goes to the kids, it's good. It's the same with time."

This guilt plays out in all sorts of contexts. But the most striking, by far, is the night shift. The next day, when I come to visit at 8:25 A.M., Clint, who has the day off, tells me that his evening duty went very well and that the kids slept until 7:30. It is only when Angie comes downstairs a few minutes later, showered, lovely, and wearing a cheerful Yoo-Hoo T-shirt, that the picture shifts: Zay, she reports, was up five times. Clint handled the first four episodes. But she got the fifth, which included a bottle, at 3:00 A.M.

"I don't think you realize," she tells Clint as we all head outside to the patio, "how many times I've been getting out of bed for the last three years."

"Sure I do."

She takes a seat and looks at him skeptically. "Even though you've been sleeping through it."

"Yup."

"How? Based on how much I complain about it?"

"No. It's not just based on how much you complain. I absolutely know how much you deal with at night, but—whether or not you'll like to hear this answer—it's because you *wanted* it that way."

Angie gives him a sheepish look. "Because of the whole cry-it-out method that I don't want to do."

"Yah."

Angie says nothing.

"After two years, you let me do it with this one"—Clint points to Eli—"and it was done in two weeks. But you didn't want to do it with this one." He gestures at Zay. "You had your method, and I let you have your method, but that method entails getting up very, *very* frequently. I didn't want to be a part of that, just like you didn't want to be part of cry-it-out."

He waits. Angie is silent. Then she makes a face. "I just don't think that you have the same response that I do to him crying. I get that internal anxiety, that physical *pain,* that *guilt. . . .*"

"I understand, it's a motherly link. You've explained that."

"So I could not be anywhere near it or hear it. Honestly, I would have to set up a cot downstairs in the office, because emotionally, I can't deal with it. . . ."

"Okay. But I think it's more like, you want me to endure the stuff that you've endured, rather than getting it done."

Angie doesn't get angry when he says this. She appears to take it quite seriously. But she's not convinced. "So last night, after the fifth time, would you have just let him cry it out?"

"No. If you were paying attention, I was increasing the amount of time I waited between each time I went in, which is how cry-it-out works."

Again, she looks at him skeptically. "Was it working?"

"Yes! I mean, I didn't have a stopwatch or anything, but yes!"

"So how come, when I started asking you questions about it, you didn't tell me what you were doing?"

"Because," says Clint, "you don't want me to do the cry-it-out method!" He looks guiltily at his toes. "At least this way you perceived it as me being really lazy and not wanting to get up. I can combat that."

It was easier, in other words, for Clint to leave Angie with the impression that he was a bum than to confess he was covertly sleep-training their child.

There was probably some passive aggression in that choice. But Clint also knew the process would fill Angie with anxiety and self-reproach, and the one thing Angie did not need in her life, clearly, was more anxiety and self-reproach. So he tried to sleep-train Zay on the sly, and then felt guilty the next day because he couldn't come clean about it. Thinking he would be judged for it, he made the executive decision that it would be better to be deemed lazy than unfeeling. But he *isn't* unfeeling.

"The way I approach this type of thing," says Clint, "is the same way I run the house. If I have $2,000, and I need to spend $1,500 on the mortgage and $400 of it on utilities, the $100 left over is going to me so that I can *maintain my sanity level.* And if I have two hours, it's the same thing. I get ten minutes, regardless."

"And I don't take that."

Clint shrugs. "If *I* don't take that ten minutes, the quality level of all the rest of it is going to go downhill really quick."

Zay is starting to fuss. Angie lets him for a second. She's thinking about this. "But lately I've been doing more for myself."

"Not as much as you could."

"I think the *want* is there," she says, "but the guilt holds me back. Like I love to go to Barnes & Noble, I love to go to movies, to be by myself. You know? And yet I don't take that. . . ."

At this point, I ask her a question: if she said, "Hey, Clint, I need an hour to go to Barnes & Noble or it's just possible that I'll go crazy . . ."

"He would say, 'Fine, go.'"

And if she said, "I really need you to take care of the kids 50 percent of the time in general?"

"I think he'd encourage me to do that too," she says, but Clint's not even here to hear her. He's gone inside with Zay.

CERTAINLY, IN SOME COUPLES, men don't do their fair share and would never even consider it, no matter how badly their circumstances require it and how far the culture has moved along. But even before meeting him, I knew Clint wasn't a slacker. In part it was because I knew that he worked long hours, both in the office and at home. In larger part, however, it was because Angie had said so herself, during her ECFE class, after she'd exorcised some of her frustration: "I mean, you've all met him! He's not a bad guy!"

But the world does not make it easy for working parents. And because of that, says Philip Cowan, "you often get all these attributions about what one person will and won't do." He and his wife hear litanies of them. "But if you have both the husband and wife in the same room," he says, "and attempt to get both sides, what gets fleshed out is how complex this is."

The thing Clint won't do, according to Angie, is the night shift. But if you ask Clint, he reframes the night-shift dispute in terms of something *Angie* won't do: sleep-train their children. More generally, he says she won't take any number of small and reasonable measures to give herself a break. "It's hard," he says, "to make Angie want things for herself."

This feeling is common. To me, it suggests that Hochschild's observation—that power struggles over who does what in a relationship aren't just about fairness but about the "giving and receiving of gratitude"—has an added layer today: guilt. Like many women, Angie feels resentment because her husband is not doing enough. But she also believes that *she* is not doing enough, and can never do enough, and that she should be doing everything all the time.

"If I were to say, 'Okay, I'll give you a break and take care of the kids 100 percent, but it's going to be *my* way,'" Clint confesses to her at one point, "I'm afraid you'll be calling the shots from the couch."

"Well," asks Angie, "what are you considering *your* way? Is your way turning on the TV and letting them do whatever, or is it taking them somewhere?"

"It's all those things," replies Clint. "If I needed to do the housework and get everything cleaned and do the dishes and make dinner and all that, there's going to be some TV time involved. I'd *occupy* them while you took your break. I'd keep them safe and engaged. But I wouldn't necessarily *entertain* them."

This may be the reason Clint believes he does 50 percent of the child care. He counts it as child care if he's doing one thing and the kids are doing another, so long as they're safe. Whereas Angie feels obliged to immerse herself completely in their world.

And Angie herself is complicit, to some degree, in this increase in her workload. Before leaving for the hospital, she fretted about the relative state of disorder she'd left for Clint. "He's going to come home to a crabby baby and a kid who hasn't napped," she said, her face bunched in a frown. "I try to get them *both* to be napping when he comes home so that he can have some free time to go to his office or go onto the computer."

It isn't only Clint who is protective of his free time, in other words. Angie protects his free time too.

"Sometimes I just assume you should *know* I'm stressed out," Angie says to him at one point. "You should see me running around, or how I'm acting. And you don't. And then I get irritated."

"And that's the part that I get irritated with," says Clint. "All you had to do was ask. *You could have just told me.*"

He is right. But asking is easier said than done. Angie experiences home as a video game, a never-ending quest to ward off flying debris. She's starting at a much higher level of stress. If you're feeling that stress, it's hard to believe that others aren't experiencing the same situation in the exact same way.

Clint may also fail to jump in and offer his time because he has mixed feelings about giving it up. The day before, for example, when Clint walked through the door, he *was* a bit miffed to discover that the kids weren't napping. "They should be sleeping right now," he told me after Angie had gone, looking slightly defeated. That pressure Angie feels to give Clint his free time is not imagined.

Though he may be unaware of it, Clint is exploiting Angie's guilt, or at the very least recognizes he benefits from it. On their mutual days off, he admits, "I'm more quick to say, 'I want to do something.' Whereas she's less quick to do that." But if he *knows* he's more aggressive about claiming his leisure time, and he *knows* that Angie is perpetually exhausted, why doesn't he yield some of his leisure time to his wife?

This is a strange moment for fatherdom. There's increasing pressure for men to be actively involved in the affairs of the home, but there's no precise standard for how much involvement is enough. In his parenting memoir *Home Game,* Michael Lewis shrewdly notes that all it takes for a couple to start fighting, really, is for them to go out to dinner with another couple whose domestic division of labor is slightly different from their own. "In these putatively private matters, people constantly reference public standards," he writes. "They don't care if they're getting a raw deal so long as everyone is getting the same deal." The problem with modern parenting, he says, is that "there are no standards and it's possible that there never again will be."

Going into fatherhood, men know they're supposed to engage. But once they're in the thick of it, many are caught off-guard, just as their wives are, by the all-consuming nature of the job. And if the standard is to do as much as their wives do . . . Lord, that bar is as high as a bird's nest. Women spend more intensive time with their children today than they have in the last fifty years.

Pamela Druckerman's solution to these excesses is to emulate the French. In *Bringing Up Bébé,* she marvels at how French parents, mothers especially, resist what William Doherty calls (in his own book about marriage) "consumer parenting," that insidious style of American child-rearing that makes it possible for a kid to lay claim to a mother's or father's attention twenty-four hours a day, seven days a week. The French, she argues, have no qualms about firmly asserting their leisure prerogatives and protecting their adult needs (like peace and quiet, for instance, or uninterrupted conversation with other adults).

That's a constructive message. But since few American women have French mothers sitting around their homes, ready to show them the way, they may do better to take their cues from a model that's more readily available: the good fathers they know. Who may in fact be their own husbands. Because odds are, these men have something valuable to teach.

Here's why: unencumbered by outsized cultural expectations about what does or doesn't constitute good parenting, and free from cultural judgments over their participation in the workforce, good fathers tend to judge themselves less harshly, bring less anguished perfectionism to parenting their children ("Sit in this Bumbo while I unload the dishwasher, would ya?"), and—at least while their kids are young—more aggressively protect their free time. None of this means they love their children any less than their wives do. None of this means they care any less about their children's fates.

Mothers may not, of course, be fully able to follow their husbands' examples. If women make more forceful claims on their husbands' free time, their husbands could well push back. This is also, admittedly, a private solution to something that in a civilized world would be a public problem, as Judith Warner argues so fiercely in *Perfect Madness*. It would be far better if the government kicked in the support that parents needed. But considering that the 2012 Republican presidential primary was briefly derailed by a debate over the legitimacy of birth control—*birth control!*—our politics hardly seem inclined in this direction. At least not yet.

For now, talking helps, especially early on. In their work, the Cowans found that the couples who had hashed out divisions of labor during pregnancy rather than after the baby came along fared much better than those who had never discussed it at all. In fact, the men who had gone through specific interventions to clarify these divisions were actually unhappy that they weren't doing *more*.

But redistributing the load is only one challenge. Another is redefining attitudes. That's what captivates me about Clint. He's so . . . forgiving of himself. Self-scrutiny and insecurity know no gender, obviously; plenty of fathers say they're terrified they're screwing things up. But I somehow think that their anguish is not the same. When I first spoke to Angie, she told me that she normally found home much harder than work—and her work involves schizophrenic and psychotic inpatients, often in the midst of violent outbreaks. Whereas Clint, who works at a desk, says he finds work more challenging. "I had to learn how to be a manager," he says. "I'm held to someone else's standard. Whereas here at home, I *am* the standard. I feel like I do it the way it should be done."

There are a lot of hardships that this pioneering generation of involved fathers has to endure. But comparing themselves to an unattainable ideal—whether it's Donna Reed in Hilldale or the "Tiger Mom" of a best-selling book—is not one of them. *I am the standard.* "Personally," says Clint, "I think having my parents separate when I was seven was the best thing that ever happened to me." Clint saw little of his father after that. "I didn't have anyone saying to me, 'This is how good you need to be.' "

Angie, meanwhile, says she never knows if she's doing things the way they ought to be done. When asked if she's a good mother, her answer is one word: "Sometimes."

She's wrong. Angie's a great mother. If she could just say, "I am the standard," maybe she would breathe.

chapter three

simple gifts

He loves to see his son's wit in the house, how it surprises him constantly, going beyond even his and his wife's knowledge and humor—the way he treats dogs on the streets, imitating their stroll, their look. He loves the fact that this boy can almost guess the wishes of dogs from the variety of expressions at a dog's disposal.

—Michael Ondaatje, *The English Patient* (1992)

THE FIRST TIME I was in a room with Sharon Bartlett, it took a long time for me to notice her, even though she was older than almost everyone around us by a factor of two. Unassuming and unadorned, she chose a place at the far end of a long table in her ECFE class, and she said nothing until the last ten minutes of the discussion, at which point it finally became clear that she was raising her three-year-old grandson, Cameron, alone. Even then, she didn't say much, but what she said was so moving that it prompted me to write her a note a few weeks later, just as I had written to Jessie and Angie and Clint, to ask if I could come visit.

"Sure," she wrote back, the very same day. "I have no problem talking about parenting my grandson. I think there are many of us grandparents parenting the children of our deceased children, and while it is not the aging plan we had in mind, it has its joys and sorrows."

In a phone call shortly after, I learned that Sharon hadn't just lost Cameron's mother. Decades earlier, she'd also lost her son, Mike, who'd taken his life at sixteen. Sharon still has another daughter to whom she's quite close—forty years old, living out of state, happy and thriving.

On a sticky morning in late July, I find myself on Sharon's doorstep. Her beautiful century-old home is in a North Minneapolis neighborhood that's largely African American, though she herself is white. She greets me at the door, coffee mug in hand and gray hair thrown loosely into a ponytail. Cam peers out from behind her legs.

"Keep playing, Cam," she tells him as she takes a seat in an oversized green chair in her living room. "I have to finish my coffee. And after, we can read five books. Go pick 'em."

Cam nods and walks over to the bookshelf. He's one of those delectable kids, all noodle arms and dewy lips. "Be sure they're different from the books we read yesterday," Sharon adds. She settles into her chair, trying to buy herself a few more moments' rest. Then she notices that Cam is squirming. She sighs and gets up again. "C'mon, Cam. Let's go to the potty. I can tell you have to go."

A couple of minutes later, they return. Sharon sinks back into her chair. Cam climbs onto the coffee table and gingerly navigates around his train collection.

"Cam?"

No answer.

"Cam Bear."

Still nothing.

"Camembert—?"

He turns around.

"Cam, that's not a step. Get down please."

And so it goes for the rest of the morning, with Sharon and Cam negotiating the vast space between a senior citizen and a preschooler.

Sharon reads Cam five stories by Richard Scarry; Cam makes a bid to play helicopters afterward, which she declines. Sharon phones her church to organize a visit to do some volunteering; Cam runs around with a towel over his head, pretending to be a ghost. Sharon hangs up and proposes they go to the water park; Cam refuses to change into his bathing suit until she counts to three, even though it's one hundred degrees outside and the city is an airborne swamp. She is patient through most of his testing and shenanigans, and when the former schoolteacher in her gets a chance to teach Cam something, she visibly comes alive. ("Look, he's making a mad face in this picture. Can I see *your* mad face?") But she also looks tired, almost painfully tired, and there are moments when her tension heaves to the surface—as when Cam accidentally smashes into her head after putting on his Crocs. "Oh, *Cam*," she says, clearly more sharply than she'd intended. "That was *not* acceptable. Say you're sorry for banging into my glasses."

"I'm sorry for banging into your glasses."

She will later tell me she feels horrible at moments like this, but I already knew. When we were chatting on the phone a few weeks earlier, she'd mentioned that she marks it on her calendar whenever she yells at Cam too much, hoping she'll one day be able to detect a pattern to her moods. Sure enough, when I go into her kitchen to refill my coffee mug this morning, I see that July 8 has been filled in with a reproachful little inscription: "Yelling Day."

All that morning, I fret for Sharon. The task of raising a toddler requires so much energy, even for the young and able-bodied. But for someone who's sixty-seven, who's already raised three children, who's living on a fixed income all by herself . . . these are hardly ideal circumstances for a parent. Most social science studies would predict that a person in Sharon's situation would be far happier without a child.

But there are things social science captures well, and there are things it does not. And one of the things it would have a hard time fully capturing, on this particular day, is what happens when we go to the splash pad.

The Manor Park Splash Pad is just a dinky patch of concrete, painted in primary colors and studded with a modest sprinkler system and some swirling gizmos. But it is heaven for a child, and on this hundred-degree day it's heaven for an adult too. The moment we arrive, Cam starts bobbing and weaving between the water jets, and Sharon, to my amazement, follows right behind him. There's a huge smile on her face, one that doesn't disappear the entire time she's there, in spite of her fatigue, her bad knees, her sixty-seven years. I think unbidden of the opening scene in the book *Immortality,* in which the narrator watches an older woman wave gaily to a lifeguard, managing for one heartbreaking instant to completely transcend her age. As Sharon stands beneath a nest of buckets, giggling while a stream of water rains down on her head, the same could be said of her. She is as unencumbered as a twenty-year-old, a picture of girlish bliss. "There is a certain part of all of us," Milan Kundera writes, "that lives outside of time."

YOUNG CHILDREN MAY BE grueling, young children may be vexing, and young children may bust and redraw the contours of their parents' professional and marital lives. But they bring joy too. Everyone knows this (hence: "bundles of joy"). But it's worth considering some of the reasons why. It's not just because they're soft and sweet and smell like perfection. They also create wormholes in time, transporting their mothers and fathers back to feelings and sensations they haven't had since they themselves were young. The dirty secret about adulthood is the sameness of it, its tireless adherence to routines and customs and norms. Small children may intensify this sense of repetition and rigidity by virtue of the new routines they establish. But they liberate their parents from their ruts too.

All of us crave liberation from those ruts. More to the point, all of us crave liberation from our adult selves, at least from time to time. I'm not just talking about the selves with public roles to play and daily obligations to meet. (We can find relief from those people simply by going on vacation, or for that matter, by pouring ourselves a stiff drink.) I'm talking about the selves who live too much in their heads rather than their bodies; who are burdened with too much knowledge about how the world works rather than excited by how it could work or should; who are afraid of being judged and not being loved. Most adults do not live in a world of forgiveness and unconditional love. Unless, that is, they have small children.

The most shameful part of adult life is how blinkered it makes us, how brittle and ungenerous in our judgments. It often takes a much bigger project to make adults look outward, to make them "boundless and unwearied in giving," as the novelist and philosopher C. S. Lewis writes in *The Four Loves.* Young children can go a long way toward yanking grown-ups out of their silly preoccupations and cramped little mazes of self-interest—not just relieving their parents of their egos, but helping them aspire to something better.

mad in the best sense of the word

After the splash pad, Sharon and Cam make their way to a playground. Cam eyes a ladder to climb. Very tempting.

"Need a boost?" Sharon asks.

"I only have two legs."

"I know you only have two legs. Give me your foot." She cups her hands, and her face turns red as he steps into them to climb the ladder. "Want to climb the next one?" she asks.

"I can't."

"Are you sure?" She picks him up and hangs him from it.

Cam looks excited and terrified. "I need to go down."

"How do you ask?"

"Please put me down?"

She does. He wanders over to another ladder with more reachable rungs. He grabs them and starts to swing. "I'm hanging from the ladder!" She keeps her distance and watches. Then he eyes a more challenging climb. "Want to hang from there too?" she asks again.

He nods.

She hoists him up. He hangs for a moment, then they move on. Sharon slips off to the side and watches. She's in no hurry. She's not looking at a watch, her cell, other moms. Nothing but Cam.

SMALL CHILDREN COME WITH a built-in paradox. The same developmental phenomena that make them so frustrating—namely, their immature prefrontal cortexes, which insist on living in the here and now—are what can also make them so freeing to be around. Most of us live by a schedule, with places to be and chores to do. But looking at Sharon, who doesn't have a day job or a husband or other small children to attend to, you begin to see how it would look if we all unshackled ourselves from the clock. Sharon doesn't bother taking her cell phone to the park when she's out with Cam (though she adores email and texting); at home she doesn't have a television. "I don't let the world intrude on me," she says. "It can only come in when I want it to." When she's with Cam, she fully surrenders to kid-time, letting the day unfold.

Few of us have that glorious flexibility all of the time. In my chapter about Jessie, the photographer and mother of three, I in fact focused on how little flexibility we have, detailing the ways in which life with a small child so often conspires against living in the permanent present. It's much easier to savor time if you're retired, as Sharon is. But not all emails need answering; sometimes deadlines exist more in our minds than they do in real time. Spending the day with Sharon is a reminder that it's easier to let time unfold than we think it is and that, in the right circumstances and frame of mind, carving out time in the permanent present is a worthy goal. It's possible to join children in their futureless worlds for even ten minutes, if that's all we've got to spare.

In Jessie's chapter, I mentioned another disadvantage to spending time in the company of creatures with immature prefrontal cortexes: they have trouble regulating their feelings, which requires an extra dose of willpower on a parent's part. But this, too, has a positive side: children lack self-consciousness. They embrace the ludicrous. They think nothing of having conversations with inanimate objects or racing buck-naked across a room.

"Usually psychologists act as if this childish uninhibitedness is a defect," writes Alison Gopnik in *The Philosophical Baby.* "And, of course, if your agenda is to figure out how to get along well in the everyday world—how to actually do things effectively—it is a defect." (Someone has to steer.) "But if your agenda is simply to explore both the actual world and all the possible worlds," she concluded, "this apparent defect may be a great asset. Pretend play is notably uninhibited."

It's amazing how many parents at ECFE talked about the joys of shedding their grown-up inhibitions, if only for a few minutes each day. For women, it tended to come out around singing and dancing: Kenya talked about watching her kid bounce and howl to Katy Perry's "Firework" in the backseat of their car; another woman talked about going to outdoor concerts. ("No one looks at you if you dance like crazy with your kid.") And then there was Jessie, who so loved her dance parties. The second time I visited her home—an evening this time, with her husband, Luke, and all three kids romping around—she showed us a new dance move Abe had done in his underpants earlier that day.

Her impersonation was dead-on, wild, ridiculous, communicated in a private shorthand that her husband seemed better than anyone in the room to understand; the two oldest kids spontaneously joined in, forming a conga line as Estelle's "American Boy" boomed in the background. When a sudden hissing sound issued from the kitchen, Luke assured Abe, who seemed concerned, that it was just the potatoes on the stove coming to a boil.

"They're screaming, 'ARRRRRRRRRRRRRRRRRRRRRGH,'" cried Luke, waggling his hands in the air. "'You're going to EAT US UP.'"

Only a four-year-old child gives you permission to imitate a despairing potato.

And these antics from Luke seemed pretty typical of the joys that the ECFE fathers described: with small kids around, they were given license to forget the gray-flannel-suit imperatives of their lives and just do what small kids do. One talked about going to the Minneapolis zoo, where he hadn't been in over fifteen years; another reveled in "watching the kids run around outside, eyeballs shining, just all teeth" (an unconscious echo, I later realized, of the refrain from *Where the Wild Things Are:* ". . . and gnashed their terrible teeth and rolled their terrible eyes . . ."). The dad who said it most succinctly put it like this: "I like that I can act like an idiot out in public."

Sometimes the transcendent joys of toddler-dom aren't about *trans*cendence at all, but about how far we can *des*cend. These joys give us a reprieve from etiquette, let us shelve our inhibitions, make it possible for our self-conscious, rule-observing selves to be tucked away. For a few blessed moments, we're streaming, uncorked ids.

It's hard to know what kind of psychic price we pay for keeping those ids bottled and sealed. Adam Phillips has always had a keen interest in this question. In one of his essays, he notes that "writers as diverse as Wordsworth and Freud, as Blake and Dickens," have all hypothesized that the turbulence and intensity we feel as young children are what ultimately give us our life force as adults. "Without this first madness," he writes, "without being able to sustain this emotional lifeline to our childhoods—to our most passionate selves—our lives can begin to feel futile."

One can argue that point, obviously. Phillips does, with himself, on the page. But he ultimately seems to conclude that there's some truth in it. He quotes the analyst Donald Winnicott: "I was sane, and through analysis and self-analysis I achieved some measure of insanity." And the route to that insanity, for Winnicott, was through the marshes of his childhood feelings. "Children, for Winnicott, are mad in the best sense of the word," Phillips writes. "For Winnicott, the question was not, 'What can we do to enable children to be sane?' but 'What we can do, if anything, to enable adults to sustain the sane madness of their young minds?' "

The tragedy, in Winnicott's and Phillips's view, would be if adults couldn't sustain this madness. Young children can at least point the way. I think about this as we leave the playground. Sharon is in a fine mood, and so is Cam. Before climbing into the car, she points happily at her toes. "Look at my feet, Cam! They're gross!" *Dirt!*

And the messiness continues, throughout the afternoon. We head off to Sharon's church, where Cam is a celebrity, clearly, and he's immediately plied with a leftover slice of someone's birthday cake. (*Chocolate icing!* Many napkins ensue.) Then it starts to rain, hard, and while Cam eats, the adults stare out the windows. The rain comes down harder, turns to hail; it's the kind of windy-wet downpour that twists umbrellas into buttercups. Cam wanders quietly over to the screen door and watches, saying nothing. When it's clear that waiting for the storm to ease up is hopeless, Sharon has an idea: *Run.*

And so we do, whooping, shrieking, all the way back to her car. Cam climbs into the backseat, and Sharon straps him in without even bothering to climb inside—she just opens the door, leans in, and lets her rump and legs get soaked in the downpour. Then she takes her seat at the steering wheel. She turns and looks at her grandson. "Pretty crazy, huh, Cam?" He nods. She nods too. "Wow," she says.

shop class as childhood

Acting like a kid isn't just about losing one's inhibitions or speaking in gibberish. Children learn about the world through doing, touching, experiencing; adults, on the other hand, tend to take in the world through their heads—reading books, watching television, swiping at touch screens. They're estranged from the world of everyday objects. Yet interacting with that world is fundamental to who we are.

This is the argument that Matthew B. Crawford makes at elegant length in his 2009 best-seller, *Shop Class as Soulcraft*. He notes that today's office workers often feel that, "despite the proliferation of contrived metrics they must meet, their job lacks objective standards of the sort provided by, for example, a carpenter's level." The information economy has made such a fetish of "knowledge work" that people no longer experience the joys afforded by knowing how to do things with their hands.

This topic was a minor theme at ECFE. "I didn't find my career particularly fulfilling," Kevin, a stay-at-home dad, told his all-male class one day. "It was just something I did. I liked it fine, but I didn't come home and say, 'Wow, I'm really glad I helped this big corporation process their data more efficiently!' "

Young children, on the other hand, offer adults the chance to engage in life's more tactile pleasures and tangible, real-world pursuits. They provide an opportunity for agency, for being able to do something and actually *see* its effect. With young children, you "make a snow slide, and it's just awesome," as one father recalled during his class. You build Lego towers, as Clint so genuinely seemed to enjoy doing. Many parents mentioned baking cookies. One mother talked about learning to bake for the first time—there's something irresistible about *thwapp-thwapping* a mound of glistening dough with a child. Overall, children make people more inclined to cook: according to a Harris Interactive poll from 2010, the overwhelming majority of Americans who cook say they do so for their families, not themselves. And what more elemental craft, with a more tangible outcome, could a human pursue?

These lost pleasures of "manual competence" are of great interest to Crawford. In his book, he argues that "the *experience* of making things and fixing things" is essential to our well-being, to our *flourishing* (to use his word), and that something vanishes "when such experiences recede from our common life." He quotes the philosopher Albert Borgmann, who makes the distinction between "things" and "devices." Things are objects we master; devices are objects that do the work for us. "The stereo as a device contrasts with the instrument as a thing," Borgmann writes. "A thing requires practice while a device invites consumption."

Now, it's true that the closets of young children today are larded with devices—devices that ding, devices that ping, devices that beep, that shine, that play music, that play videos, that respond to a simple touch. But early childhood is also one of the few times when we as a culture still emphasize the supremacy of—and mastery over—*things.* We buy our kids hammers to bang and necklaces to bead; we give them finger paints to smear and plastic instruments to play; we sit on the floor and lay acres of railroad track, build towers of Tinker Toys, make flowers out of pipe cleaners. When a child is born, there's always a relative who goes off and buys that child a tool set, thinking he or she ought to know how to use it. In preschool all children learn music, all children do arts and crafts, all children use blocks, play catch, dance. Parents are often surprised to discover that their children are just as interested in using screwdrivers to open the battery compartments of their various devices as they are in the devices themselves. They still view devices as *things.* You can take them apart and put them back together. Children still have their hands on the world.

Perhaps the explanation for this is a simple one, anchored in a basic developmental reality: early childhood is when we first gain control of our bodies and develop our motor skills. But in some ways, that's the point. Toddlers and preschoolers acquire knowledge in ways that are *inseparable* from their physical experiences. This is the time when it's easiest to see what we human beings may truly be—"*inherently* instrumental, or pragmatically oriented, all the way down," as Crawford suggests. By spending time with young children—building forts and baking cakes, whacking baseballs and making sand castles—we're afforded, in some respects, the opportunity to be our most human. This is who we are. Creatures who use tools, creatures who create, creatures who build.

philosophy

When her eight-month-old was asleep and her two older kids were watching TV in the next room, I asked Jessie what she loved most about parenting. I thought her answer would be her dance parties. And she mentioned them. "But on a bigger scale," she said, "I love watching my kids learn how to figure things out on their own. It's how an explorer must feel."

It's a little clichéd to say that small children are always changing. What's so pleasurable about Gopnik's *The Philosophical Baby* is that she describes these changes neuroscientifically, and at times even quantifies them. Part of what's so startling about the minds of babies and young children, for instance, turns out to be a simple function of volume and frequency: their brains are so plastic that their intellectual inventory turns over every few months, making their learning curve a true spectacle to behold. "Imagine," Gopnik suggests, "that your most basic beliefs would be entirely transformed between 2009 and 2010, and then again by 2012."

Children remind us just how much of our implicit knowledge, which hums inaudibly in the background all day long, is stuff we once had to *learn*. They climb into the bathtub partially clothed, put half-eaten bananas in the refrigerator, use toys in ways the manufacturer never intended. (*So you want to mix those paints rather than make pictures with them? Lay stickers on top of one another rather than lay them out side by side? Use dominos as blocks, cars as flying machines, tutus as bridal veils? Knock yourself out!*) No one has yet told them otherwise. To children, the whole universe is a controlled experiment.

And that's just the practical stuff. One woman at ECFE told me her daughter asked whether she'd always be a girl—not because she wished to be a boy, but because she didn't know whether gender is a fixed trait or a mutable one. A man told his all-dad ECFE class that his son, earlier that day, turned to him after staring out the window and said, "Maybe when we're squirrels, we'll be up in that tree." (He didn't know whether our roles—this time in the entire animal kingdom—are fixed or mutable.) Mihaly Csikszentmihalyi, author of *Flow,* shared a similar moment from his young parenting life when he told me about the time he took one of his sons to the beach: "He saw some bathers, swimmers, coming out from the water, and he sort of froze, saying, 'Look, Water People!' It was so . . ." Csikszentmihalyi didn't complete the sentence. But the word he was searching for, I think, was "logical," because he continued: "I thought, *Yes, I could see how, if you've never seen them, they seem like extraterrestrials.*" Of course! Swimmers: Some otherworldly species that makes its home in the sea.

MOST ADULTS CONSIDER PHILOSOPHY a luxury. But philosophy, it turns out, is what children do naturally, and when they do, they take us back to that remote and almost unimaginably luxurious time when we ourselves still asked loads of questions that had no point. In fact, according to Gareth B. Matthews, author of *The Philosophy of Childhood*, asking pointless questions is the true specialty of children, especially between the ages of three and seven, because the instinct hasn't yet been drummed out of them: "Once children become well settled into school," he ruefully observes, "they learn that only 'useful' questioning is expected of them." (Which recalls Edmund Burke's alleged observation about studying law: "It sharpens the mind by narrowing it.")

René Descartes once said that a person has to start over in order to do philosophy properly. "That is hard for adults," writes Matthews, who taught philosophy at the University of Massachusetts at Amherst for more than three decades. "It is unnecessary for children." Children have nothing to unlearn. Matthews gives a perfect example—the concept of time—and quotes St. Augustine: "What, then, is time? Provided that no one asks me, I know. But if I want to explain it to a questioner, I am baffled."

Parents, like St. Augustine, are often baffled when their children ask questions about something so basic as time. But they are often delighted by them too. There's something decadent, and at the same time intellectually pleasing, about entertaining such fundamental questions. "A couple nights ago," said an ECFE father, "Graham and I were snuggling, and he goes, 'Dad? What's water?' " Graham was two and a half. He knew what water was, obviously. But his question was: "Yeah, but what *is* water?" There was an audible ripple of enthusiasm around the table. *What is water? Yes!* "And I'm like"—his dad clapped his hands, theatrically warming to the subject—" 'Well, there's hydrogen, there's oxygen.' . . . It was awesome."

After class, this dad told me that Graham's follow-up question was even wilder: "Can you *break* water?" "Because I'd told him that if you put the hydrogen and oxygen together," he explained, "it makes water. So he wanted to know if you could break it up."

I heard a number of similar questions that week. ("Why are people mean?" was one of my favorites. Also: "Is there only this place, the place with the sky?") Matthews's books are filled with such examples too. He recounts telling a classroom of adults about a classic question asked by a child—"Papa, how can we be sure that everything is not a dream?"—and a mother replying that her three-year-old daughter had just asked the modern-day analogue: "Mama, are we 'live' or are we on video?" Even more striking than his examples of existential questions, arguably, are his examples of children stumbling onto questions of ethics. He mentions a child who, after seeing his dying grandfather, asked his mother on the car ride home if the elderly are ever shot when they're ready to die. His mother, obviously startled, answered no; that would be troublesome for the police. (A strange answer, perhaps, but any parent who's been in such a situation can identify with the slightly panicked instinct to keep replies concrete and the discussion brief.) The boy, then four, responded, "Maybe they could just do it with medicine."

"In important part," writes Matthews, "philosophy is an adult attempt to deal with the genuinely baffling questions of childhood." Many adults enjoy pondering philosophical questions if given the chance. But they have little excuse to do so in their everyday lives, until they have children. Then they've got a chance, at least for a few years, to contemplate—and perhaps reconsider—why the world around them is what it is. He quotes Bertrand Russell, who said of philosophy: "If it cannot *answer* so many questions as we could wish, [it] has at least the power of *asking* questions which increase the interest of the world, and show the strangeness and wonder lying just below the surface even in the commonest things of daily life." Kids have an uncanny knack for asking those questions. And the questions, as far as Matthews is concerned, are the true revelation, not the answers.

love

When Sharon first laid eyes on Michelle, the baby who would one day grow up to be Cam's mother, she was just five months old and weighed eight pounds. "Failure to thrive" was the term the agency had used when they put her in Sharon's care—Michelle's biological mother, whose intelligence was well below normal, had clearly neglected her, though to what extent couldn't be fully determined. Sharon received her happily, lovingly, as she had the nine other foster children she had cared for, but Michelle made a deeper impression somehow. Maybe it was because she was so young; maybe it was because she was small and vulnerable and sweet. Whatever the reason, Sharon and her two biological children were smitten—"we loved her to smithereens"—and that love only deepened over time. Five years later, Sharon found herself standing in front of a judge, submitting the last of the paperwork, to make the adoption official.

Sharon didn't start noticing behavioral problems until Michelle was nine years old. Michelle had an IQ of only 75, which may have contributed to her defiance; the frustration of coping with so many cognitive difficulties can easily spill into social difficulties, as parents of severely learning-impaired children can attest. Or perhaps Michelle's behavior was the result of those five first crucial months of neglect, which for all Sharon knew involved unspeakable abuse; perhaps it was scribbled into a few obscure strands of her DNA. Michelle also began acting out around the same time her brother, Mike, died, which Sharon assumes was no coincidence. Everyone in the family was suffering terribly back then. Mike's suicide had a wretched effect on everyone.

But whatever the origins of Michelle's unruliness, Sharon suddenly had a wildly oppositional child on her hands, a girl who never finished high school and would frequently run off to live with various boyfriends. The psychiatrists called it "attachment failure." In practical terms, it meant something much more basic to Sharon.

"It took a long, long time for Michelle to believe I loved her," she says, as we settle into her living room for the afternoon and Cam naps upstairs. "A lot of the opposition and the testing was from the *utter disbelief* that someone could love her." Throughout Michelle's late teens and twenties, she would leave home and come back, leave and come back, and each time she disappeared it'd be for months at a clip, with Sharon never knowing whether Michelle was dead or alive.

People close to Sharon were amazed at her forbearance. "They would always ask me, 'Why do you do this to yourself, taking her back in? She's just going to break your heart again,'" says Sharon. "And I would say, 'Well, because I'm her mother.'"

At this point, I confess to Sharon that I too am amazed that she found the strength to cope. Wasn't it hard to funnel so much love into someone who *resisted* love and attachment?

She shrugs. "It's the whole bonding thing, I guess."

Right. But she's describing a child who *didn't* bond.

"But I bonded to *her,*" she says. "She might not have bonded back. But I bonded to Michelle." That was enough. "I can't tell you why I loved her," she says. "But I loved her. I always did."

ON PAGE 1 OF *The Four Loves,* C. S. Lewis makes a distinction between what he calls Gift-love and Need-love. "The typical example of Gift-love," he writes, "would be that love which moves a man to work and plan and save for the future well-being of his family which he will die without sharing or seeing; of the second, that which sends a lonely or frightened child to its mother's arms."

In my conversations with parents of young children, it's Need-love that often most bedazzles, and with good reason: there's nothing like it. To be adored unconditionally, to be hoisted on a plinth and held above reproach, is rare for most adults, no matter how much they're loved by their spouses or cherished by friends.

"Maybe it's egocentric to say this," said a woman in Angie's ECFE class, "but you're their whole world, at this age. I love that. . . ."

". . . and maybe that's why," finished the woman next to her, "you don't want them to grow up."

Many adults need the people they love too. But in young children, love is almost *indistinguishable* from need, which makes their adoration especially powerful. Because they live in the present, forgiveness comes easily to them; they haven't yet formed the mental machinery to hold grudges. ("When I apologize, she's instantly okay," another woman told her class. "She's like, 'Yeah, Mom, that's fine.'") Toddlers and preschoolers do not smolder, do not hoard a satchel of grievances, do not love conditionally. They love. That's that.

Yet it's Gift-love, and not Need-love, that parents spoke about with more passion. Need-love comes from children. Gift-love is something parents give away. It's a far less cozy arrangement. Gift-love can be difficult to muster too, contrary to what so many cheerful books about new parenthood contend. It does not come instantaneously to all parents the moment they're handed their new baby in the nursery. Rather, it blooms with time. Alison Gopnik makes this distinction with a perfect aphorism in *The Philosophical Baby*: "It's not so much that we care for children because we love them," she wrote, "as that we love them because we care for them."

That's the kind of love that Sharon had for Michelle. Through nurturing her as a child, day in and day out, Sharon came to love her, to bond with her, and to wish stubbornly to protect her, no matter how hard Michelle pushed back as a teen or an adult.

This is not to suggest that parents have cornered the market on Gift-love, or that they're better at it than nonparents. There are plenty of times when parents feel themselves loving conditionally, qualifiedly; there are moments when they discover, to their horror, that they have trouble loving their children at all. Sharon is acutely aware of her past shortcomings as a young mother, decades before Cam came into her life. She and her husband divorced early, which left her broke, sometimes incapable of providing her kids with so much as bologna sandwiches for lunch. She remembers missing crucial milestones—first steps, first words—because she was beating the pavement, trying to find teaching work. After Mike died, she struggled with a terrible depression, which she knows had consequences, rendering her absent while still home, and making her daughters suffer a double loss, not just of their brother but of their mother too.

"I don't want to say I was a *bad* parent back then," says Sharon. "But I was such a needy person, trying to raise these kids and not giving them what they needed." She has compassion for her younger self. She also knows her life would have been easier if, as she puts it, "there'd been more societal helps." But the struggles of those years defined her inner life and her subsequent choices. "Knowing that I had not been able to make my kids feel like their world was safe, and that help was available when they needed it, drives a lot of what I do for other people now," says Sharon. "And it's driving *everything* that I do with Cam."

What Cam has given her, many years later, is another chance to be her best self. That's what children can do in the end: give us that shot, even if we so frequently and disastrously fall short of the mark. It takes effort to love generously. "There is something in each of us that cannot be naturally loved," writes Lewis toward the end of *The Four Loves*. "Every child is sometimes infuriating; most children are not infrequently odious." But on the days we're at our best, our very finest, we're able to overlook these imperfections and love our children with only their best interests in mind. “There are many ways to approach that ideal and to care for others—ways that don’t involve children,” Gopnik notes. “Still, caring for children is an awfully fast and efficient way to experience at least a little saintliness.”

ABOUT NINETY MINUTES HAVE passed. Cam suddenly appears at the top of the stairs. “Cameron!” shouts Sharon. “You’re awake! Come here, you!” He races down the steps, runs over to his grandmother, and gives her a giant embrace. She pat-pat-pats his bottom. “Did you have a good nap?”

He nods. It’s hard not to be struck by how much happier Sharon looks now than she did this morning. I ask if she was aware that she couldn’t wipe the grin off her face when we were at the splash pad earlier.

She smiles. “No. But I love water. That might be part of it.” She gives it a bit more thought. “I love kids, I love water—they all came together there. And I knew Cameron was having fun. It made me happy to see you laughing in the water, didn’t it?” she says, turning to Cam. “You were laaaaaughing. You were drinking the water!” She stares right into his face. “It made me happy.”

So Sharon strives, in her own way, for a little bit of saintliness. In fact, she’s a religious person, intensely active in her Catholic church. She’s not really sure about God—“I’m more of a Jesus freak than a God freak,” she likes to say—but she believes in the Gospels and social justice, framing her philosophy in very simple terms: “You should take care of people who don’t have what they need.”

That’s what her life with Michelle was about: trying to give her what she needed. “And I was getting *something* back,” Sharon says, after thinking it over. “She was receptive to my love.” Sharon considered that a gift. Michelle was the kind of child who found it hard to trust anyone, much less accept affection. It meant something to Sharon that Michelle learned to accept hers. "It felt good to love her," she says. "Also, it was *my promise* to love her. I'm very strong on promise. I can't be *who I am* if I take on an obligation and I don't carry through with it." She recalls the day she stood before the judge, finalizing the adoption. "He looked at me and said, 'You understand, you cannot bring her back.'" She laughs, pretending to reply. "*Yes, I get it. It's for life.*" And that was the point. "At the time I made my choice, I had other options," she says. "But once I made my choice, I stripped myself of those options. That was it, right there."

And now she's made the same commitment to Cam. At thirty-two, Michelle came back to Sharon after another months-long disappearance, telling her she was pregnant. What Michelle didn't know was that she also had advanced-stage cervical cancer, because she'd never gone for any internal exams during her pregnancy and the cancer didn't show up on any of her sonograms. Cam was born at just twenty-eight weeks, a frail little thing whose prospects for survival were hardly better than his mother's. Michelle held on for nine more months. She was in a lot of pain toward the end, incapable of being touched, incapable of holding her son. The one wish she made clear was that Sharon should care for him. And so Sharon does, every day, imperfectly but fiercely, with Cam now occupying the same room in which his mother spent her final months.

"I'm in it for life now," says Sharon. "And sometimes I do think to myself, *Would he be better off with two parents, with some young people in his life?*" She can do the math. She's sixty-seven; he's three. "And then I'm like, *Possibly. But where would I find them, and how would I know for sure?* And if I can't know for sure, and I can't find them, he's got me."

Albert Einstein is supposed to have said that there are two ways to lead a life: one in which we act as if nothing is a miracle, and the other in which we act as if everything is. Sharon directs her gaze across the living room. Cam has discreetly settled into an armchair, Richard Scarry's *A Day at the Airport* sprawled across his lap. "When Cam was born," she says, "he weighed three pounds and was fourteen inches long. And now, a mere three years later, he sits with his legs crossed, reading a book. How can you not be incredulous and amazed?" She stares at him for a while. We both do. "How does this happen?" she asks. "We start as *nothing*. And today, just look at him. Is he in charge of his earth or what?"

chapter four

concerted cultivation

Profound must be the depths of the affection that will induce a man to save money for others to spend. —Edward Sandford Martin, *The Luxury of Children and Some Other Luxuries* (1904)

IT WAS LAURA ANNE Day's idea that I come to Cub Scouts sign-ups with her. It wasn't the plan when I originally phoned. I had no plans. I knew only the barest details about her life: she was thirty-five, divorced, and a mother of two boys; she worked full-time as a scheduler for a lawyer; and she lived in West University Place, an affluent, autonomous city within the city of Houston, a place where striver parents typically lead busy, ambitious lives.

Then she told me she was a Scout mom. Cub Scouts are a huge deal in Houston. Laura Anne can tell you sixteen different reasons why they're a big deal, but none of them has anything to do with why she suggested I tag along with her this evening. Rather, Laura Anne believed that Cub Scout sign-ups were the best place in the neighborhood to witness the real-time theatrics of parents trying to set their children's fall schedules. As she's driving us there in her Toyota Highlander, her sunglasses crowning her blond bob and her khaki Scout shirt tucked neatly into her jeans, she warns me that the air will be filled with arias of conflict and overcommitment from the moment we walk through the door. "You'll see," she says, locking the car door.

She's right. The very second parents arrive at the West University United Methodist Church, they start to shower the Scout leaders with questions: How many times per week do I need to attend these meetings? Is there any flexibility when it comes to my kids' participation? Because . . .

"Because my son has at least an hour of homework per night," says a mother, looking at her son, "and piano and soccer, and my younger one has T-ball . . ."

"Because he has Skype lessons for Indian classical music once a week," says another woman, "and voice lessons twice a week, and piano and soccer and language lessons on the weekend—Sanskrit on Saturday and Hindi on Sunday . . ."

"Because," says a father just a few moments after that, neatly summing things up, "like everyone here, we are seriously overscheduled."

Randy, a podiatrist by day and the Cubmaster to whom these questions are metronomically directed this evening, nods understandingly each time. "Right. The answer is, there are two meetings per month, one with the den and one with the pack. . . ."

As he elaborates, the father grimaces. "Okay. I'll have to consult." He looks at his son. "Because, naturally, he's got football Tuesday nights." *Naturally,* because this is Texas, and everyone plays football in Texas. And *naturally* in the ironic sense too, because the Cub Scouts' once-monthly pack meetings are also on Tuesday nights. "And I'm his younger brother's soccer coach on Tuesday nights too," says the dad. Meaning he's doubly unavailable to take his kid to Scouts. "Okay. . . ." Yet here he is, trying valiantly to figure out a way to make it all work. "Maybe he just won't go to football practice once a month," he says, letting out a long, resigned *pffffff.* "Or maybe . . ." He trails off. "Maybe we'll just clone ourselves."

the overscheduled parent

It was William Doherty, the professor of family social science at the University of Minnesota and adviser to ECFE, who, in 1999, first coined the phrase "overscheduled kids," thus contributing the perfect term to describe the sudden proliferation of play dates and extracurricular activities on children's agendas, as if they'd all suddenly acquired chiefs of staff. Overscheduling has earned more than its share of critics, who fear it makes kids anxious and robs them of the glories of imaginative idling and unstructured play. But few critics think to ask what kind of harm such overzealous planning might be doing to children's *parents.* Yes, those parents are usually the ones responsible for the family schedules and are therefore complicit in the problem. But it's worth considering what particular forces could be driving mothers and fathers to such extravagant lengths.

Because they *are* extravagant. Behind every overscheduled child is a mother or father filling out forms, hustling from T-ball to ice-skating to chess lessons, and, in many instances, going through the same paces the child is, from learning the violin Suzuki-style to co-building miniature replicas of Reliant Stadium for school. As one mother put it to me: "I planned a career that would allow me to work part-time because I wanted to be a stay-at-home mom. And I am never at home."

The sociologist Annette Lareau was one of the first to take an in-depth look at this controlled pandemonium, capturing it in energetic detail in *Unequal Childhoods,* which became a classic the instant it was published in 2003. Looking at a dozen families—four of them middle-class, four of them working-class, and four of them poor—she couldn't help but notice some crucial differences in parenting styles.

Poor and working-class parents did not try to direct every aspect of their kids' lives. She called their approach the "accomplishment of natural growth." The style of middle-class parents, on the other hand, was something altogether different—so different she coined a term for it: "concerted cultivation."

"Concerted cultivation," Lareau writes, "places intense labor demands on busy parents, exhausts children, and emphasizes the development of individualism, at times at the expense of the development of the notion of the family group." Throughout the book she looks at middle-class parents with a sense of both empathy and mild, bewildered awe, but what seems to surprise her most is the psychologically insidious nature of middle-class parents' engagement with their children's lives. She uses a family called the Marshalls to make this point most vividly. "Unlike in working-class and poor families, where children are granted autonomy to make their own way in organizations," she writes, "in the Marshall family, most aspects of the children's lives are subject to their mother's *ongoing* scrutiny [emphasis hers]." Like which gymnastics program her daughter should attend, for instance. "The decision regarding gymnastics," she writes, "seemed to weigh more on Ms. Marshall than on any other member of the family." It was as if her daughter's future depended on where she learned her back flips and cartwheels.

This is one of the reasons I've come to Houston and its surrounding suburbs. The area is one of the country's unofficial capitals of concerted cultivation, though it may not be associated with it in quite the same way as New York or Cambridge or Beverly Hills. The place has all the ingredients. It's got a thriving middle class. It's obsessed with sports. It has a large population of adults who do research for a living—at the Texas Medical Center, at Baylor and Rice and the University of Houston, at a host of energy companies. And it's exploding with families with children under the age of eighteen, according to the 2010 census.

After-school baseball isn't just Little League here. It's the *right* Little League team and a private batting tutor; advanced kids do club-level tournament baseball, for which "there aren't even tryouts," says Laura Anne. "You have to be *asked*." Bela Karolyi, the former coach to Nadia Comaneci and Mary Lou Retton and Kerri Strug, runs a gymnastics camp sixty miles outside of Houston. Some kids start football before they can read. "Stephen went to football camp over the summer," says Monique Brown, another mother I met, "and those parents were like, 'We gotta get these kids protein shakes and muscle milk!' As if their kids were going to go pro. At *seven*."

Summers can be even worse, because there are full days to fill. When I first started phoning parents in Texas, it was June, and I spoke to the mother of two boys, ages eleven and thirteen, both gifted in math and science. Most parents know that summer camp today isn't the summer camp of yore, filled with tetherball and relay races and inedible Jell-O snacks. Camp has become a week-to-week series of immersion courses, each designed to nurture strengths and open minds. But even in light of these new, talent-optimizing developments, the recitation this woman gave of her sons' summer options was truly Homeric, amounting to a nerd's paradise for the kids but a carpool Hades for her and her husband. There was one computer camp to learn Java, another to learn C++, and another to learn Visual Basic; still another camp taught "Game modding," teaching aspiring hackers off-menu ways to supercharge their video games. Among the offerings at the Natural Museum of Science were Chemistry Camp, Space Camp, Dinosaur Camp, and Physics Camp (which included the Star Warriors Academy and Council); the American Robotics Academy offered camps on "crazy action contraptions" and "Leonardo da Vinci machines." The brainiest seventh-graders could participate in a three-week program allowing them to take college-level courses, from architecture to neuroscience. More eccentric kids could do "duct tape creations," which, just as it sounds, was devoted to making improbable objects out of the sticky stuff, including cell-phone cases and flip-flops.

But the simplest way to tell this story of mushrooming children's activities and escalating parental involvement in them is through numbers. Back in 1965, when women had not yet become a regular presence in the workplace, mothers *still* spent 3.7 fewer hours per week on child care than they did in 2008, according to the American Time Use Survey, even though women in 2008 were working almost three times as many paid hours. Fathers, meanwhile, spent over three times as many hours with their children in 2008 as they did in 1965.

So how to account for this steady trend toward a more exhaustive—and exhausting—style of parenting?

One explanation is straightforward: we are having fewer children, which means we have more time for each child. But other explanations are more subtle. We live in a nation of sprawl, which means our homes are farther apart, making it tempting to enroll our children in organized activities just to give them a chance to socialize. (This, I should add, is especially true of the Houston area, looped with ribbons of twelve-lane highways that rattle with SUVs the size of ice-cream trucks.) We are also surrounded by a tantalizing buffet of electronic media, whose lures parents fear ought to be counterprogrammed with more constructive agendas. Today's parents harbor extra—and not always rational—concerns about children's safety, which makes them more inclined to control their time. And we live in a nation of women who work, a fact that still generates discomfort and ambivalence, resulting in stricter imperatives for parents, mothers especially, to spend more of their nonworking hours with their children to compensate for all that time away.

Perhaps most fundamentally, though, hyperparenting reflects a new sense of confusion and anxiety about the future. Today it is the unimpeachable conviction of the middle class that children ought to be perfected and refined in order to ready them for the world ahead. But our higgledy-piggledy efforts to do so, often contradictory and even promiscuous in nature, suggest we're a bit flummoxed about what that entails and what our exact role *is* in this important matter. What, precisely, are we preparing our children *for?* How, as mothers and fathers, are we supposed to prepare them for it? Have parents always operated this blindly? Or were the roles of parents and children more clearly—and simply—defined in the past?

The answers to these questions may seem obvious. But they're more complicated than one might think, and they lie at the heart of some of the most obdurate challenges of the middle parenting years. It is during these years that the strengths and weaknesses of children start to reveal themselves in all their contours, as do their preferences and tastes. The things that once were quirks, modestly irritating habits, or joyful tendencies start to harden into full-blown characterological traits. The whole person starts to show. Middle-class parents perceive these years as a crucial stretch of time, one in which they can either do something or nothing to coax out their children's best selves. But they're uncertain how to do it, and they're uncertain whether their goals for their children are even possible. "When the kids were younger," says Leslie Schulze, a mom in the Houston suburbs you'll be meeting shortly, "the question was, 'Am I starting cereal at the right time?' Now, it's 'Am I making the right decision for my child?' "

Indeed, how would she even know if she were?

WE ARE ACCUSTOMED TO thinking about concerted cultivation as the neurotic product of coastal overprivilege or Lone Star State–size ambition. But as I was reading Annette Lareau, the first two people I thought of were a pair of women I'd met through ECFE in St. Paul. Both were members of a group of moms who referred to themselves as "The Committee," because they all knew one another from committees associated with their children's activities. "All the women in The Committee," wrote one of them to me in an email, "are intelligent, caring, fun, and dynamic women who are very close to being completely burnt out."

The next day I met with the author of the email, Marta Shore, in a café. Marta is a statistics professor and mother of two. She told me she was so fried that she decided this year to limit her parental involvement to just one activity per child.

Why? I asked.

She looked confused for a moment. "Because if I didn't impose the limit of one per child," she finally said, "it'd be eighteen."

But that wasn't my question. I was asking why she felt the need to do anything *at all.* She was already driving her nine-year-old daughter to swim lessons and Hebrew school and piano (or, at the very least, coordinating the carpools that got her there), and she was already a leader of her daughter's Girl Scout troop. She was investigating judo for her too, as well as art lessons—not that she could afford these activities, not that there was time—but she thought she ought to at least consider these options because her daughter had expressed an interest in them. She was also doing advocacy work on behalf of ECFE, because she and her three-year-old were still attending classes.

Yet Marta couldn't envision a world in which she *didn't* do all these things. When I asked her why she was holding herself to such a high standard (all on a modest budget, and all while working full-time), she no more understood the question than if I'd asked why she persisted in breathing.

The following day, I met her friend Chrissy Snider, another

woman on "The Committee." Chrissy stays home with her four kids, and she too told me she was "on a sabbatical year," though she still held positions on the church council and child ministry team. I asked about her kids' extracurricular schedules. Here was her response, transcribed verbatim:

> Eddie will do two sports this summer. And he'll have stuff during the day—swimming five days a week for five weeks. And he'll have an art class too, but they're with his older brothers, so I only have one drop-off. But then he'll have T-ball and soccer, which are going to be a different schedule than Henry, who's on traveling soccer and rec league baseball, and Ian, who's just in rec league baseball. And then they both have tutoring for reading. Henry does piano and cello. He does cello at school, and he does private piano lessons. And he's weighing which one, because it's a financial thing, unless we win the lottery. He wants to do both, because this is his niche. And Ian does violin. That's at school, but I have to be at his lessons, because it's Suzuki.

Her youngest, Megan, was not yet old enough for extracurriculars. She was only two.

Neither of these women is flush with means. Their kids go to public schools. Chrissy and her husband and four children live in a 1,300-square-foot house—not exactly a palace. Each extracurricular choice they make means forfeiting some other pleasure. (Several times during our conversation, Marta mentioned the high cost of date nights with her husband, between the babysitting and the evening activity itself.) But this is what they do. It's what other parents around them do. It's also what they *read* they should be doing—this is what happens when you raise children amid the dust rings of the information age. "I read, 'Girls who do sports are less likely to do drugs or get pregnant,' " said Marta, "and my response was, *Oh no, if she doesn't do soccer at four, she'll never do a team sport.*"

It took a while for it to occur to her that perhaps it didn't matter, that her daughter would be just fine.

the rise of the useless child

Today most middle-class parents take for granted that Marta's and Chrissy's way is the natural way. As far as children are concerned, there is no such thing as excess. If improving their children's lives means running themselves ragged—and *thinking* themselves ragged—then so be it. Parents will do it. Their children deserve nothing less.

Yet adults did not always take this indulgent view of children. Before the nineteenth century, they were distinctly unsentimental about them, regarding childhood "as a time of deficiency and incompleteness," according to historian Steven Mintz. Rarely, he writes, did parents "refer to their children with nostalgia or fondness." It was not uncommon for the New England colonists to call their newborns "it" or "the little stranger," and no extra measures were taken to protect these little intruders from harm. "Children suffered burns from candles or open hearths, fell into rivers and wells, ingested poisons, broke bones, swallowed pins, and stuffed nutshells up their noses," writes Mintz. Nor did grown-ups try to shield children from the more brutal emotional realities of life: "As early as possible, ministers admonished children to reflect on death, and their sermons contained graphic descriptions of hell and the horrors of eternal damnation."

These quotes all come from Mintz's meticulously detailed *Huck's Raft*, a chronicle of American childhood from the nation's beginnings to the present day. To any parent who isn't a professional historian, the book is a revelation. Any good history provides useful context for present-day conventions and belief systems, but reading about the history of childhood is especially startling, because we tend to think of our beliefs about children as instinctive, and therefore as unchanging, *irreducible.* Yet according to Mintz, Americans hardly started with the notion that children are vulnerable and adorable innocents, although the idea was not completely unfamiliar: in the eighteenth century, the philosopher Jean-Jacques Rousseau argued that children are pure and spontaneous creatures, free of inhibition and guile, and John Locke argued in the seventeenth century that children are born a blank slate, amenable to a parent's guidance. But it wasn't until the early nineteenth century that adults began to think of children as precious. That's when the high chair first made its appearance, literally signifying children's newfound, elevated role (they'd earned themselves a place at the table, so to speak); the first advice literature on child-rearing appeared; and the United States saw the beginnings of public schools. Institutions protecting child welfare began to spring up, like children's hospitals and orphanages. A thought revolution had begun.

For most children, however, this revolution didn't translate into new privileges. Kids were too valuable economically. In the early nineteenth century, the Industrial Revolution created a massive demand for child labor. In small towns, children headed off to work in the mills and mines that had begun to dot the landscape; in cities they flooded the street trades and factory floors. As agriculture began to be commercialized, child labor on farms became especially valuable; in fact, children *already* were integral to the farm economy, yanking weeds by the time they were five, according to Mintz, and harvesting crops by the time they were eight. In the late nineteenth century, children were more likely to earn money for their families than their mothers were, and the wages of teenage boys often exceeded those of their dads.

Not until the Progressive Era—loosely defined by most scholars as the period between 1890 and 1920—did adults finally make an organized, concerted effort to ban child labor. Change was still slow.

Reformers generally made exceptions for farm work, because it was considered character-building; during World War II, restrictions on child labor had to be relaxed because so many young men were overseas. But the end of the war was the tipping point. Childhood as we think of it today—long and sheltered, devoted almost entirely to education and emotional growth—became standard for American children. Only adults worked full-time. Children did not; they could not. In fact, parents began giving *them* money: that strange custom we all know as "the allowance" officially began. The primary job of a child became his or her schooling. "The useful labor of the nineteenth century child," writes Viviana Zelizer in *Pricing the Priceless Child*, "was replaced by educational work for the useless child." Homework replaced actual work. Which had value, certainly. *But not to the family.* "While child labor had served the household economy," notes Zelizer, "child work would benefit *primarily the child* [emphasis mine]."

Children, in a funny way, became the first true members of the information economy. Schoolwork, which corresponded little to the life skills needed to run the house, became their area of greatest expertise. Academics and sports. Modern childhood had begun.

As I noted in my introduction, Zelizer found a memorable, five-word phrase to describe this historic transformation. Children had become "economically worthless but emotionally priceless."

CHILDREN BENEFITED A GREAT deal from this new sentimentality. Being considered precious and irreplaceable gave them far more power in the family hierarchy than they had ever had when they were contributing to the family till. Some sociologists went so far as to argue that their newfound sacred status stood the traditional family structure on its head. Writing in *Fortune* magazine in 1953, urbanist William H. Whyte described the postwar United States as a "filiarchy," or culture in which kids run the show, at one point even calling children's influence "dictatorial." (He'd go on to write the best-selling *Organization Man* in 1956.) The moment children stopped working for adults, everyone became confused about who was in charge.

This inversion has had even more pronounced behavioral consequences for the middle class today. "Middle class children," writes Lareau in *Unequal Childhoods,* "argue with their parents, complain about their parents' incompetence, and disparage parents' decisions." This was not true of the children she followed in poor and working-class homes, where they would "respond promptly and wordlessly to directives from adults." Lower-income parents, she noticed, give orders and directives. Middle-class parents give choices and negotiate.

Children sense their solicitousness. The kind of lip discouraged and punished by parents in other eras is something middle-class parents now reward. While all children were once told, more or less, to know their place, only the less affluent—who lack power to begin with—are told to think this way today. Middle-class children, on the other hand, are told that they are fully empowered. In the long run, this attitude may or may not serve them well, because they then enter the world with the sense that no power structure is too formidable to defy or outmaneuver. But one thing is immediately clear: this attitude is not very good for *parents.* "The very same skills parents encourage in their children," writes Lareau, "can and do lead children to challenge, and even reject, parental authority."

The newly emboldened child may help explain the appeal of the Boy Scouts to so many mothers and fathers. The Scouts teach order. They teach respect. They aren't the only institution to do this, of course; religious institutions do too. But the Scouts put *parents* in charge, not strangers dressed in robes. As Randy, the podiatrist and Cubmaster, told me: "The kids see you as a role model, and then they start to act out the things we're doing automatically." He thought this over for a moment. "It's very *nice* to have your kids be polite. Or to go to a restaurant and have them act properly."

HAD THE CHILD'S ROLE been the only role to change within the family, that alone would have been a significant historical development. But industrialization and modernization inevitably changed the role of parents too. As time went on, mothers and fathers also lost their traditional function in the family economy. Before the Industrial Revolution, parents provided educational, vocational, and religious instruction to their children; they also tended to them if they got sick, helped make their clothes, and supplied the food on the table. But with industrialization, these jobs were gradually, one by one, outsourced to non-family members or entire institutions, to the point that the idea of the "family economy" practically ceased to exist. The sole job of parents became the financial and physical security of their children.

Ever since, every debate we have had about the role of parents—whether they should be laissez-faire or interventionist "Tiger Moms," attachment-oriented or partial to the rigors of tough love—can be traced back to the paring down of mothers' and fathers' traditional roles. Today, we are far less clear about what "parenting" entails. We know what it *doesn't* entail: teaching kids mathematics and geography and literature (schools do that); providing them with medical treatment (pediatricians); sewing them dresses and trousers (factories abroad, whose wares are then distributed by Old Navy); growing them food (factory farms, whose goods are then distributed by supermarkets); giving them vocational training (two-year colleges, classes, videos). What parenting *does* involve, however, is much harder to define. The sole area of agreement for almost all middle-class parents—whether they make their children practice the violin for three hours a day or exert no pressure on them at all—is that whatever they are doing is for the child's sake, and the child's alone. Parents no longer raise children for the family's sake or that of the broader world.

As parents, we sometimes mistakenly assume that things were always this way. They weren't. The modern family is just that—modern—and all of our places in it are quite new. Unless we keep in mind how new our lives as parents are, and how unusual and ahistorical, we won't see that the world we live in, as mothers and fathers, is still under construction. Modern childhood was invented less than *seventy* years ago—the length of a catnap, in historical terms.

the globalized, optimized child

"Did you find the place all right?" Leslie Schulze, forty and lovely and dressed impeccably well, is greeting me at the door of her redbrick home in Sugar Land, an upscale suburb southwest of Houston. Though it has five bedrooms and 5,200 square feet to roam around in—she and her husband bought the place roughly a decade ago, for $350,000—her house is by no means outsized for her neighborhood. Her well-tended block is studded with homes of equally ample capacity (two family rooms, cathedral-like kitchens, en suite bathrooms for all), and the drive to her place, along Palm Royale Boulevard, features larger houses still, a few on the order of 14,000 square feet. (I first mistook them for country club facilities, until I noticed they were popping up one after another, in some kind of *Dynasty* version of tract housing.) Shortly after my visit with Leslie, I would be told repeatedly by other Sugar Land residents that wealthy Indian doctors owned a number of these sprawling estates, a fact that isn't entirely confirmable, obviously, and to some ears might strike a gratuitously racial note. But I soon learned that this attunement to demographic change had a larger context. In 1990 Sugar Land was 79 percent white. Today it is 44.4 percent white and 35.3 percent Asian. The politics of the district were and still are conservative—it was represented by Tom DeLay, the House Majority Leader, until his 2006 resignation—but the recent influx of high-achieving immigrants from India and China has changed the face of the area and, in the view of many white Sugar Land residents, directly affected their lives by rewriting the standards of academic excellence in the local schools.

"If a Caucasian family here is going to push its kids," Leslie tells me, "they're more likely to push them in sports, like putting their children in year-round swim team until their shoulders pop out. Whereas the Asian families put more of an emphasis on academics."

True to her own distinctions, Leslie insists that her children do at least one activity per week that's physical. Her thirteen-year-old daughter is on the volleyball team, which meets every day; her ten-year-old son does baseball twice weekly, karate twice weekly, and flag football twice a week in the off-season. Academic accomplishment is undeniably important to Leslie too. She speaks with pride about her daughter being an honors student, and the previous summer she hired a tutor for her son because a statewide test for third-graders was fast approaching. But you can hear, in her voice, a certain incredulity that it's come to this. "I kept thinking to myself," she says, "*Why am I hiring a tutor for a kid who's still in second grade?*"

It used to be that Leslie's approach worked. If you moved to an upper-middle-class enclave like Sugar Land and encouraged your children to do a team sport and get good grades, odds were that your kids would one day find a place in the state's fine system of universities, and then march into the marketplace with a degree and a fistful of contacts. Their place in the world was reasonably assured. Leslie's husband grew up in Texas and went to the University of Houston, where he met Leslie; they are passionate about their alma mater and post lots of Facebook updates about football games they've just attended (*Go Coogs!*)

But recently, there's been a rather striking assault on the long-cherished order of things. In 1997, the state legislature passed House Bill 588, more commonly known around the state as the "Top 10 Percent Rule." The bill essentially guarantees college admission (though not the means to pay for it) to the state-funded universities for all kids who graduate in the top 10 percent of their high school class. The measure drew support not just from poor blacks and Hispanics, who appreciated that their children would be measured relative to the accomplishments of their peers (rather than those whose parents could afford homes in better school districts), but from poor rural white Texans too. But the legislation has had interesting consequences for suburbs like Leslie's. Where once Texas A&M, the University of Texas at Austin, and the University of Houston took a disproportionate number of kids from the state's best public schools, located in places like Sugar Land, today these institutions must hew to the numbers. And the kids who graduate in that top 10 percent, in Leslie's view, are the children of Tiger Moms. (It just so happened that Amy Chua's book *Battle Hymn of the Tiger Mother* had come out a few months before my visit.)

"Whites in my daughter's middle school," Leslie tells me, "are less than 50 percent"—in fact, they make up just 31 percent, according to the school's own data—"while Asian and Indian are the most" (53 percent). "So it's very academically competitive now. That's a big concern in terms of college admissions. When my daughter finishes eighth grade, she'll have two high school credits. But there will be kids with five and six." So by the time high school graduation comes around, says Leslie, "she'll be lucky to be in the top quarter of her class."

The distinctions Leslie is making may seem a bit oversimplified, and she may be making them with less hedging self-consciousness than some of her peers. But she is describing a common sentiment among her cohort. The rules for preparing kids for this networked, outsourced, multicultural century are brand-new. She brings up a parent open house she attended the first week of school the previous year, when her daughter was starting seventh grade. "All the Asian parents raised their hands," she recalls, "and asked about the Duke University program." Leslie had never heard of it. "And a number of the other Asian and Indian parents start nodding. But it wasn't on my radar. I walked out saying, *I have no idea what they're talking about.*"

Later, she learned. The Duke University Talent Identification Program is a national organization that, among other things, invites seventh-graders to take the SAT or ACT, ordinarily taken in grade 11. If a child scores over a certain threshold, he or she may participate in various summer and academic-year curricula held throughout the Southeast and Midwest. When Leslie learned that her daughter, a smart girl, qualified, she "contacted the school counselor and said, 'What are the benefits? Is it something colleges will look at?'" Reasonable questions, she figured. But what she got were maddeningly vague answers. "Their response," she says, "was, 'It's up to you.'" The program was not going to show up on her daughter's transcript, and it was not going to help her get into Duke University one day, though it had Duke's name on it. "So I encouraged her to a degree," says Leslie, "but I said it was up to her."

Her daughter chose not to go. "And in hindsight," says Leslie, "I feel like I should have pushed. But I was reluctant. She had enough activities. She was all pre-AP. I didn't want to stress her out."

THOUGH MOST PEOPLE DON'T realize it, Margaret Mead had a lot to say about American parenting. Her observations appear in one of her lesser-known works, *And Keep Your Powder Dry,* originally written in 1942. Parts of the book seem quaint today. But the pages about the way we raise our children and the anxieties we associate with them may as well have been written last week. It's not an accident that Mead wrote the book at roughly the same moment children were losing their traditional roles and the modern, sheltered childhood was invented. Better than almost any social critic, and almost without meaning to, Mead documented what happens to parents when their primary obligation to their children is to cultivate them.

As an anthropologist, Mead was familiar with many different kinds of family structures and child-rearing philosophies. What struck her as different about American parents was that they didn't know what particular goal they were steering their children toward. If you were an English aristocrat, your aim was to raise your child to be another aristocrat. If you were a rice farmer in India, your aim was to train your child to farm the same land. If you were a blacksmith in Bali, you raised your son to be a blacksmith too. Whether you lived in old-world Europe or a developing nation or a preliterate society, you, as a parent, were a custodian of old traditions, not an inventor of new ones. "In other societies, where parents were bringing up children for their own way of life," she writes, "the job was reasonably clear. As you sat, so you taught your children to sit." It didn't matter if you were a clumsy teacher to your child. Behind "the ignorance and ineptness of any individual," she observes, "lay the sureness of folkways." A parent's job was simply to maintain them.

But Americans, Mead immediately noticed, were a special case. They didn't have old folkways to rely on. They didn't have a "way of life" that they were raising their children for. The whole promise of America—its very appeal, its very strength—was that its citizens were not hidebound by tradition or rigid, immutable social structures. Americans were—are—free to invent and reinvent themselves with every generation. Sons and daughters do different things from their fathers and mothers, in different ways, and in different places. "The American parent expects his child to leave him," she writes, "leave him physically, go to another town, another state; leave him in terms of occupation, embrace a different calling, learn a different skill; leave him socially, travel if possible with a different crowd."

In many ways, this is an incomparably wonderful thing. But it leaves mothers and fathers with few guideposts to direct their children. Following Mead to her logical conclusion, Americans are trying to ready their sons and daughters for a life that will look nothing like the lives they themselves lead. The word Mead uses to characterize an American father's relationship to his sons is "autumnal"—a lovely word, and even more resonant in today's world of snappity-pop technological change—because he is preparing his sons to surpass him. "In a very short while," she writes, "they will be operating gadgets which he does not understand and cockily talking a language to which he has no clue."

Uncertainty, Mead notes, makes even the brightest and most competent mothers and fathers vulnerable. In America, uncertainty starts the moment their children are born. New parents in the United States, Mead observes, are willing to try almost any new fad or craze for their baby's sake. "We find new schools of education, new schools of diet, new schools of human relations, sprung up like mushrooms, new, untried, rank like skunk cabbages in early spring," she writes. "And we find serious, educated people following their dictates." Which is why attachment parenting is considered de rigueur one year and overbearing three years later. And why cry-it-out is all the rage one moment and then, after a couple of seasons, considered cruel. And why organic, home-milled purees suddenly supplant jars of Gerber's, though an entire generation has done just fine on Gerber's and even gone on to write books, run companies, and do Nobel Prize–winning science. Uncertainty is why parents buy Baby Einstein products, though there's no evidence that they do anything to alter the cognitive trajectory of a child's life, and explains why a friend—an extremely bright and reasonable man—asked me, with the straightest of faces and finest of intentions, why I wasn't teaching my son sign language when he was small. ("Because my mother didn't teach me to sign," I told him, "and look! I'm smart and pay my taxes and everything.") It is why, in any bookstore, you will find parenting advice guides by the hundreds, by the thousands, many of them starkly contradicting one another. *There is no folk wisdom.* "In old static cultures," Mead writes, "one can find a standard of behavior—a child will be judged a baby until it can walk, or a small child until it has lost its teeth or learned to pass kava or take the cows to pasture . . . but in America, there is no such fixed standard—there are only this year's babies."

Uncertainty is also why parents of school-age children pack their schedules cheek-by-jowl with what they hope will be enriching and life-readying activities. As far as a child is concerned, Mead writes, "all one can do is to make him strong and well-equipped to go prospecting for himself." How one equips children to go prospecting for themselves has never been more unclear, but teaching them chess for strategic thinking, enrolling them in AP classes, encouraging them to do sports so that they'll learn the virtues of teamwork and perseverance and resilience—these all seem like they might help. We are no longer training our children for a trade or a guild or a place on the farm. As Mintz notes, American society since World War II has uniquely assumed "that all young people should follow a single, unitary path to adulthood"—namely, the staircase of kindergarten through grade 12, followed by a trip to college, if they're lucky enough to be among the middle class. This singular path turns all peers into potential competitors, all measured by the same metrics (the SAT, GPAs, AP scores, a slate of extracurricular involvements). The only thing parents can do in such an environment is help give their children a boost above the fray, cultivating them with the same assiduousness with which children once cultivated the family fields. Nora Ephron put it this way in 2006:

"Parenting [is] not simply about raising a child, it [is] about transforming a child, force-feeding it like a *foie gras* goose."

WHEN MEAD WROTE *AND Keep Your Powder Dry,* the United States was a fairly homogenous nation. It would be another twenty-three years before the Immigration and Nationality Act of 1965, which eased restrictions on Asian, Latin American, and African immigrants, was signed into law by President Lyndon Johnson. But once it passed, we became the most diverse nation on the planet, and Mead's observations took on an extra resonance. They acquired more meaning still as the economy globalized and the world's borders began to dissolve. Not only were parents training their children for a life that was radically different from their own, but they were training their children for a life that would, potentially, be conducted in a different language. It was this anxiety that made parents encourage their children to study Japanese during the 1980s, and it is this same anxiety that drives a certain kind of parent to enroll their toddlers in Mandarin classes today (and explains the sudden appearance, in 2007, of Nickelodeon's *Ni Hao, Kai-Lan*). It's as if parents, uncertain about what future to prepare their kids for, are trying to prepare them for any and every *possible* future.

Leslie expressed this anxiety with singular directness. But I heard it from other parents too, framed in different ways. Like Chrissy, the Minnesota "Committee" mom who had outlined her four kids' formidable extracurricular schedules for me. After an ECFE class one day, I asked: where did all of her internal parenting pressures come from?

"I think they come from a variety of places," she told me, "but one of the things that set me off was Tom Friedman." She was referring to the *New York Times* columnist, who frequently writes about globalization. I should add that we were not talking about anything like globalization at the time. "His book *The World Is Flat,*" she continued, "and the idea that Chinese families and Indian families are working so hard to raise these kids, and they were going to take all of the jobs." She shook her head. "After reading that, I was like Ho-ly Buckets!" Her children's future jobs could just as easily be snatched away by a student in Delhi as in Denver.

But didn't she still feel like her four kids had lots of natural advantages? I asked. They were white middle-class Americans, and they went to good magnet schools in St. Paul. . . .

"I *do* think my children have a lot of natural advantages," she told me. "But the world is changing, and we've seen so much change just in my lifetime. You just don't know what's going to happen in ten, fifteen years."

Chrissy didn't have the sureness of folkways, as Mead would have said, to guide her parenting.

the globalized, optimized child, part ii

Leslie may believe that her cohort's emphasis on sports isn't as useful for college admissions as geometry might be. But not all parents feel that way. Our beliefs about what will get our children ahead are idiosyncratic, really, based on a murky combination of conjecture and personal experience. Tiger parenting can assume many shapes and forms.

I see this very clearly one late summer afternoon in Missouri City, another suburb of Houston, as I sit at a local park with Steve Brown. Missouri City is a mite different from Sugar Land: more African American (42 percent) and not quite as well-to-do (though still well off). What Missouri City does share with Sugar Land, however, at least at the community level, is a mania for sports. Steve, in particular, is *really* gung-ho about sports. We're here to watch his seven-year-old son play soccer.

"Y'all scrimmaging today?" Steve asks as he sets up two mesh chairs. Steve and his wife are both black and both Southerners, and they live in a neighborhood that is pretty but more modest than Leslie's, with houses running about $150,000 each. Looking at this particular park, though, you wouldn't sense much of a difference: It has more beautifully maintained sports fields than I can count.

"No," says his son. "Saturday." He heads into the field. We take our seats in the shade.

I ask Steve whether soccer was his idea or his son's.

"It was . . . a family idea," he says. He's handsome, personable, intense; his shoulders are taut and become even tauter when he watches his son play. "See, as a kid, I'd gone and played in all different sports. I didn't settle into what I wanted to do"—tennis—"until I was in high school." He still carries the aesthetic with him. He's dressed from head to toe in tennis whites. "So I kind of want to expose him to different things. He's at the age where he *could* do football. And we put him in a camp, over the summer, just to get a sense. But soccer's one of those sports where the skill sets are transferable."

Steve likes to say that in his family, when he was growing up, "sports were king." A tennis scholarship made his college education affordable; a basketball scholarship made one affordable to his father.

Steve's son swings by for a swig of water. "Where *is* everyone?" Steve asks him, looking at a half-empty field.

"I don't know." His son runs back out.

I know something else, though, about Steve's son: he's not necessarily the competitive type. I can see it, and Steve's wife, Monique, has basically told me as much. So I ask Steve: what if his son turns out to be the kind of kid who likes to paint?

He laughs. "Maybe he'll be," he says. "But at least through ten or so, he's going to do this." Then he explains. "My brother wasn't really into athletics. We're twelve years apart. And Dad didn't push him. He didn't coach him the way he coached me." He speaks with a certain delicacy next, a certain carefulness. "And I don't know how that impacted his personal growth, not being pushed. But we're clearly two different people, me and my brother."

Steve's brother has done just fine for himself, actually. But Steve—people *know* Steve. If he wants to argue that his college commitment to competitive sports has given him an edge in life, it's hard to quarrel with him. He has his own public affairs firm in Houston (he lobbied on behalf of the Top 10 Percent Rule, in fact); he's the chairman of the Democratic Party in Fort Bend County. I had no idea he had such a high profile in the community when I first contacted him. I found him through the parent-teacher organization of Palmer Elementary School, one of the most diverse in the area; his wife, Monique, and several other parents in this chapter are on the board. It just so happened he was a well-known local guy.

So sports and success, I say, are intertwined, as far as he's concerned.

"Uh-huh," he says. "You have a will to succeed that most people don't sense. I think that's one of the things you carry over into your real life. A strong will and desire. And ambition."

He looks again into the field. "Where *is* everybody?" he repeats.

"So for a time, we'll keep doing this," he says, keeping an eye on his son. "It gives him structure, it teaches him discipline and teamwork—certain things that are transferable to real life. To make sure he has a foundation. He may get to the point where he doesn't want to compete. And that'll be fine. I think athletics, though, will still be a part of our lives."

So that he and his son will still have something in common? Or so that he stands a better chance of earning a scholarship for college?

"Well, an athletic scholarship wouldn't hurt," says Steve, and laughs. "That's part of it." He reflects for a moment. "That *is* part of it," he suddenly says, much more soberly. "We're competitors in my family. That's how we were raised."

Then he makes a point I wasn't expecting, but perhaps should have. "I do believe that soccer's going to be the basketball of the next generation," says Steve. "So it'll be a pretty good sport to excel in. More so than when I was growing up."

This is the age of globalization, in other words, and soccer is the most popular sport on the globe. So if you're the Tiger Mom of sports—which Steve cheerfully admits he is—and if you want your kid to be competitive in these changing times—which Steve most certainly does—then soccer is the game to play.

IT IS FREQUENTLY PRESUMED that concerted cultivation is the province of the most narcissistic parents. In some cases, this is true. (Hence the term "trophy child," a devastating new addition to the parenting lexicon.) All of us have met the parent who natters on with false humility about the over-accomplished wonderfulness of his or her children. But there's also a more charitable way to interpret this mad rush to cultivate middle-class kids. One could simply say it's a legitimate fear-response, a reasonable and deeply internalized reaction to a shrinking economic pie.

When my parents bought their first house in 1974, it cost them $76,000. They managed the down payment with the help of my two grandfathers. Even correcting for inflation, that kind of money is bird-seed today. To buy the same house would cost *three times* the adjusted amount, and it is doubtful that men of my grandfathers' occupations—a Brooklyn hospital administrator and a movie-theater projectionist in Queens (the man to whom patrons shouted "Focus!" and "Louder!" if the picture and volume weren't exactly to their liking)—could contribute in the same meaningful way. (And they contributed, I should add, almost equally.) Today's dollars don't stretch as far, and middle-class families don't have the same amounts socked away. On the eve of the recession, the average household debt exceeded its disposable income by 34 percent.

In *Perfect Madness*, Judith Warner rehearses the details of middle-class economic decline in painful detail: while a record number of American households are worth more than $5 million, the adjusted wages of middle-income families haven't budged since the 1970s; mortgages for the average home have made up a larger and larger proportion of household income; health care costs are now brutally high (even for families with private insurance, averaging 9 percent of their income, according to the Obama White House's Middle Class Task Force). Men especially have suffered demoralizing setbacks in their earning potential in the last few decades. Between 1980 and 2009, college-educated men between the ages of thirty-five and forty-four (prime working years, at least in theory) have seen their wages rise at less than half the rate of productivity. And women, it turns out, pay a steep economic price for being mothers: according to Shelley Correll, a Stanford sociologist who looks at gender inequities in the labor force, the wage gap between mothers and childless women who are otherwise equally qualified is now greater than the wage gap between women and men generally.

But perhaps the most terrifying number to today's parents comes from the US Department of Agriculture, which estimates that a child born in 2010 will cost a middle-income family $295,560 to raise. If the family is high-income, the number is $490,830; if the family is low-income, it's $212,370. These price tags do not include college tuition, which is rising at a rate that vastly outstrips inflation. The average annual cost of a private, four-year college came to more than $32,600 in 2010, and a public college, nearly $16,000.

Under these constrained circumstances, it's hardly surprising that today's middle class looks at their children and fears that they will have little purchase in the world by the time they grow up. Any well-meaning parent would do what they could to assure their children an edge, and therefore a slightly brighter fate.

Ironically, this panic has played out in some of the most flamboyant ways among the middle-class families with the most means—namely, the *upper*-middle class, who seem to be the most startled by these economic changes and most threatened by losing their prerogatives. In 2005, the economists Peter Kuhn and Fernando Lozano wrote a paper noting that we're now at an unusual economic juncture: the best-paid men in the American workforce are far more apt to put in long work hours than those who make the least (specifically, the top 20 percent versus the bottom 20 percent). The reverse was almost always the case throughout the twentieth century. And it's not the monetary rewards that motivate them to work long hours, these economists speculate; it's more likely that they hope to distinguish themselves in an uncertain job climate and, by doing so, land an added measure of job security at an insecure time. The opportunity cost of *not* working a lot is too great.

Annette Lareau and her colleagues have shown that there's a parenting analogue to this phenomenon. Mothers who have gone to college, they found, enroll their children in far more organized activities—in large part for the same reasons the best-paid men work so much. These mothers, too, believe that the opportunity cost of not enrolling their children in loads of extracurriculars is too great. It's the problematic psychology of any arms race: the participants would love not to play, but not playing, in their minds, is the same as falling behind.

the irreproachable mom

"Do you have my sunglasses?" asks Benjamin Shou, thirteen, as he opens the car's rear door. It's 2:45 P.M., and his school has just let out. He dumps his backpack into the empty seat next to him.

His mother, Lan Zhang, answers in Chinese. She and Ben always speak in Chinese. But because I'm around, she switches to English. "Do you want to eat something? Fruit? Water? Gatorade?"

"No," says Ben. "How long am I skating?"

"Two hours."

"What?" He frowns. He's not a frowny sort of kid usually. He tends to be self-assured, can-do, all smiles. He was the one who coordinated this meeting with his mother, though his mother speaks extremely good English. "I have so much homework," he says. "I have a math test tomorrow."

Lan, who's prone to smiles herself, and not very big on formality (she's wearing jeans and no jewelry, unless you count the hair scrunchy on her wrist), gives him a curious look in the rearview mirror. "What? It should be Friday."

"No. Tomorrow. And a vocab test too."

"So how many hours of homework do you have today?"

"Three."

Now it's her turn to frown.

"Do you want to take a nap? You have twenty-five minutes." That's how far away it is to Sugar Land Ice and Sports Center.

Ben shakes his head.

"So what else. Any news from school?"

"Yes. I have so much homework!" He grins.

"That's the third time you've told me! Maybe you can talk to your coach." And she returns his grin in the rearview mirror. "Poor Ben. Everyone pushes you."

IF ONE WERE INCLINED toward bald-faced stereotyping, Ben looks, to all outward appearances, like the incomparable product of a Tiger Mom. He attended a prestigious magnet school in Houston until sixth grade; today he goes to St. John's, one of the city's most storied private schools. He started figure skating at six and today trains six days a week, two of them in Sugar Land; at the age of twelve he placed fourteenth in the juvenile division of the 2011 US National Junior Figure Skating Championship. He does Boy Scouts on Tuesday nights. He takes piano lessons on Sundays, for which he practices about half an hour each day. He gets very good grades. How could an indefatigable Tiger Mother *not* be behind at least some of this?

Lan, Ben's mother, is indeed indefatigable. But she also perfectly demonstrates the nuance that's missing from any polemic about parenting (or any subject, for that matter). Lan's affect is extremely gentle. There's an anxiety to her, and it's not of the sort that results in aggression; it's much closer to ambivalence and vulnerability. A lot of Ben's endeavors weren't her idea. He's the one who became obsessed with ice skating, after seeing skaters one Christmas in the nearby mall. He's the one who expressed an interest in playing the piano, after seeing a neighbor's child play.

And for the record, Lan doesn't send Ben to Kumon, the afterschool enrichment program that first appeared in Japan in the fifties and now has dozens of centers in the Houston area alone. She tried once, when Ben was in the fourth grade, but he hated it and she didn't push. "I'll be honest," she says as we sit in a gallery above the rink and watch Ben skate. "I don't *like* Kumon. I come from China, and the system there is so strict. I hated that. I want Ben to grow up like a normal kid."

In fact, Ben is currently only one grade ahead in math, just like Leslie's daughter, while many of his friends are two. To Lan, it's enough. The idea that excellence no longer suffices and super-excellence has become the new standard clearly gives her trouble too. She says she flinches every time she hears friends or fellow parents comparing results on standardized tests. "I don't want Ben to hear that," she says.

"I only want him to know that if he wants to achieve, he has to work hard. But I don't want to push him."

Did she read Amy Chua's *Battle Hymn of the Tiger Mother?*

"Only pieces," says Lan. "It was kind of dramatic. And from my experience, even if moms act like Tiger Moms, they would never *say* they are Tiger Moms."

One characteristic that Lan does share with Tiger mothers, however, is that she has given up quite a lot for the sake of her child. One wouldn't have necessarily predicted this trajectory for her: she comes from a family of well-known Chinese artists and intellectuals, and she grew up with a belief in expressing herself. She learned the violin from her father and began publishing poems and articles at just eight years old. After graduating from Beijing Normal University, she became a reporter and editor in China. It was only when the MD Anderson Cancer Center offered her husband, Jiang, a postdoctoral degree that she came to the United States.

Here, she worked at a Chinese-language newspaper for a while. But then she had Ben, and everything changed. She stayed home with him for the first four years, and when he went off to preschool, she decided a reporter's life wasn't for her, because it involved too many late evenings. Instead, she took some courses in biology. Now she works at Texas Children's Hospital, doing research on gene therapy.

Lan finds the work challenging. But it's not her first love. Writing is. What her job provides is compact hours so that she can be home with Ben at the end of the day. Every morning, she drives him to school. On the weekends, she or her husband drives him to his piano lesson. She drives Ben to the rink on weekends too, and during the weekdays she sometimes takes him to a different rink early in the mornings, before school starts. There's also his scouting activities on weekends and Tuesday nights.

I mention that she must spend an awful lot of time in the car.

"It's horrible," she says. "If we lived in Sugar Land, it'd be much easier, but then it'd be too far to drive to work every day." She looks thoughtfully at the ice. "What if one day he tells me, 'I don't want to skate; I want to play basketball.' It happened to other kids I know. Kids. Anything can happen."

It seems unlikely. This summer Ben spent seven weeks in Colorado Springs, training daily at a huge rink from six o'clock in the morning until six at night. He'd then come home from practice and do an online math course, at the insistence of his father. Lan went for five of those weeks, using up two years' worth of vacation time in order to do it. Ben's father went for three weeks too. They stayed in an apartment two miles away,

Well, I say, she could tell Ben he *can't* quit.

"I would never do that," she says. "I tell him, 'If you love it, I can spend time, money, and energy to support you; you just have to do your best.'"

Yet in some ways these expenditures—time, money, and energy—are the least of it for Lan. What's scariest, she says, is all the emotional capital she's invested in her son's life.

"Parents want their kids perfect," she says. "But they can't be. And you see that in skating." She points. Ben is in the midst of a very graceful program, as best as I can tell. "See, this jump he just did?" I did. It was impossible to miss. "It's a double axel," she says, "but he's still not consistent on this. He's very happy if he lands it, but if he doesn't, he's very sad. And I feel sad too. It's very hard to keep your emotions under control."

A woman with a long ponytail starts following Ben around the rink, her wrists clasped behind her back. Lan explains that she's his trainer, Shannon. We stare for a while. He is truly magnificent. "He has a magic for skating," says Lan, clearly thinking the same thing. She's finally allowed herself to relax, and the pleasure, the pride, is legible on her face. *I can't believe this boy is mine.*

"When you have your logic, you think, *Oh, skating is only skating. The kids love it,*" she says. "But before competition, it's not only about fun. And you can't stand outside. You become so involved."

Again, she stares at the ice. "My life has three parts," she says. "One is job. One is Benjamin. And the other is after ten, when I do my writing and editing. But sometimes I'm too tired."

And if she had more time to herself, would she pursue her writing?

"Yes," she says. "There are many books I want to write." She's already had two collections of articles published in China. "But I'm supposed to publish three books together. It's a series. I can't finish. I don't have time."

I ask if Ben has ever read her books. She shakes her head. She's supposed to know his world, not the other way around. Even if Ben wanted to read her books—and if ever there were a kid who would, it'd be Ben—he couldn't. Unlike so many other parents in her cohort, Lan never made her son learn to read Chinese.

IN CHAPTER 2, I talked about how mothers bear the brunt of most child-rearing tasks. This imbalance extends well beyond the toddler years. In 2008, Lareau and her colleague Elliot Weininger analyzed two different data sets, each made up of families with elementary school–age children, and concluded that "women's lives are much more heavily intertwined with children's organized activities than are fathers' lives"—a fact that surprised them, given that such a high volume of kids' after-school activities tend to be sports. (One reason, they hypothesize, is that "fathers might be involved as coaches with *one* activity, while mothers did the rest.") Mothers in their samples continued to assume the roles of scheduler, logistics coordinator, and family nag, just as they had in the earlier years. The mothers were also the ones who continued to fret and bear the psychological load:

> It was mothers who signed their children up for activities, figured out how to transport children to their practices, reminded them to rehearse their instruments, pressed their clothes for recitals or their uniforms, and found out where the traveling team would be playing the next Sunday.

Perhaps more significantly, Lareau and Weininger's study suggests that "at least some employed mothers face a tradeoff between time devoted to paid work and time devoted to facilitating their children's leisure"—which explains why a woman like Lan would give up on a reporting career to find more flexible work in a lab. Organized leisure, unfortunately, is *not* flexible (Scouts meet only on Tuesday evenings), and it's not always predictable either ("Wait, you won sectionals? So where are we driving this weekend?"). A week speckled with scheduled activities amounts to a week of what the authors call "pressure points," or non-negotiable and time-sensitive demands that fall disproportionately on mothers. "Their time-use patterns," they write, "are more frenetic than those of their husbands."

These time-use patterns are happening at a peculiar inflection point in our culture. On the one hand, the number of Americans who believe "women should return to their traditional roles in society" is just 18 percent, according to a 2012 survey by the Pew Research Center. On the other hand, the number of Americans who believe children would be better off if their mothers stayed home full-time is now 51 percent, also according to Pew, from a survey taken just one year later.

Our expectations of mothers, in other words, seem to have increased as our attitudes toward women in the workplace have liberalized.

On the face of it, these disparate trends seem to contradict one another. But one could also argue that they're related: as a culture, we may be more ambivalent about women's ubiquity in the workplace—and the nonfamilial child-care arrangements that result—than we care to admit.

History certainly suggests as much. In the past, at just the moments women had gained some measure of education or independence, the pendulum often took a wild swing backward, with the culture suddenly churning out the unambiguous message that women ought to be seated back at the hearth. A number of books have made this argument over the years, but Sharon Hays's *The Cultural Contradictions of Motherhood,* published in 1996, still ranks among the most cogent to me. In her view, whenever the free market threatens to invade the sanctity of the home, women feel greater pressure to engage in "intensive mothering." Even the best-intentioned child-care experts have made their female readers feel this way. Hays points to T. Berry Brazelton, a best-selling author of her day, who declares in his book *Working and Caring* (1985) that, "in the workplace, a woman . . . must be efficient. But an efficient woman could be the worst kind of mother for her children. For a home, a woman must be flexible, warm, and concerned." And this example seems positively quaint today in light of the newfound enthusiasm for attachment parenting, which, while appealing in many ways, requires a formidable time investment on the part of the mother, who theoretically is not allowed to leave her child's side until that child is three. For a family requiring two incomes, this arrangement is hardly practical, nor is it practical for a woman who has different priorities for how she spends her time.

These are just two recent examples of the enduring link between women's independence and calls for more attentive mothering. In her 2003 book *Raising America,* Ann Hulbert makes a virtuoso survey of twentieth-century child-rearing practices and finds many examples predating the present day. At the turn of the twentieth century, a time when more and more women were college-bound, child-rearing experts proclaimed that higher education was the perfect preparation for motherhood because children were endlessly interesting subjects of study and therefore infinitely worthy subjects of cultivation. (Thanks to his mother's college education, wrote a prominent thinker of the day, "no boy of hers will get to that sorrowful age when he feels that he knows a great deal more than his mother.") In the 1920s, just as women were cropping their hair short and exercising their newly won right to vote, researchers were urging mothers to return home and pay more attention to the new, emerging field of child development. (From a 1925 article in the *New York Times:* "By some strange cosmic alchemy, the same economic and social forces which have broken up the old-fashioned home and sent women into the world of business and pleasure on much the same terms as men, upsetting the manners and morals of the race, have now distilled a new interest in the business of being a parent.") Even the year the word "parent" first gained popularity as a verb is interesting. It was 1970. At the same watershed moment that women were yanking off their aprons, taking the Pill, and fighting for the Equal Rights Amendment, the word "parent" entered common usage as something one could *do* all day long.

But perhaps the starkest example of this backlash phenomenon was the Eisenhower era, which formed the backdrop to Betty Friedan's landmark manifesto of second-wave feminism, *The Feminine Mystique,* published in 1963. World War II proved to be a time of great flowering for women: they married later, for obvious reasons, and took over domestic jobs long held by men (most famously as defense-industry workers); they also made contributions closer to the front lines as nurses and in the Women's Army Corps. But there was a retrenchment during the fifties. Though women continued to work, they didn't enter the job market with the same ambitions as women ten years earlier. The average age of first marriage for women dropped to twenty in 1950, "the youngest in the history of this country," writes Friedan, "the youngest in any of the countries of the Western world, almost as young as it used to be in the so-called underdeveloped countries."

Happily, many of the problems Friedan wrote about in *The Feminine Mystique* feel outdated today. But that doesn't mean we aren't in the midst of a new backlash against women. It's just of a different sort.

Back in the fifties, women felt great pressure to keep an impeccable *house*. The words "Occupation: Housewife," which women wrote on census forms if they didn't work outside the home, form a leitmotif in Friedan's book. Women felt pressure to be fine mothers too, of course, but the symbol of it all, and the locus of their efforts, was the home. Dinners had to be splendid and punctual; beds had to be made; floors had to be buffed to a high shine. Never mind that single-minded devotion to these pursuits often left women feeling hollow and unfulfilled, an emptiness Friedan famously called "the problem that has no name." The upkeep of a fine home was a woman's work, and if she found it unrewarding, well, she simply had to turn the prism another thirty degrees to see that she'd been mistaken: it *was* an important job, and by no means beneath her. Madison Avenue was in the business of telling her so. One of the most revelatory chapters of Friedan's book features internal research documents she'd secretly obtained from a consultant, which she quotes to skillful effect:

> One of the ways that the housewife raises her own prestige as a cleaner of her home is through the use of specialized products for specialized tasks . . . when she uses one product for washing clothes, a second for dishes, a third for walls, a fourth for floors, a fifth for venetian blinds, etc., rather than an all-purpose cleaner, she feels less like an unskilled laborer, more like an engineer, an expert.

This was Madison Avenue's solution to the problem that had no name. If women felt restless, or like their jobs as housewives were beneath their educational attainments, the answer was to counter that their jobs most certainly did require educated people—women were domestic scientists.

Today women have abandoned this form of domestic science, spending almost half as much time on housework as they did in Friedan's day (17.5 hours per week, to be precise, versus nearly 32 hours per week in 1965). But they have become domestic scientists in another way: they're now parenting experts, and they spend more time with their children than their mothers ever did. It was a woman in Minnesota who clarified this shift for me. She pointed out that her mother called herself a *house*wife. She, on the other hand, called herself a stay-at-home *mom*. The change in nomenclature reflects the shift in cultural emphasis: the pressures on women have gone from keeping an immaculate *house* to being an irreproachable *mom*.

And the market today, still hoping to appeal to women's professional instincts, offers the same differentiation in baby products for mothers that it offered in cleaning products for housewives sixty years ago. Back in the fifties, women were told to master the differences between oven cleaners and floor wax and special sprays for wood; today they're told to master the differences between toys that hone problem-solving skills and those that encourage imaginative play. This subtle shift in language suggests that playing with one's child is not really play but a job, just as keeping house once was. Buy Buy Baby is today's equivalent of the 1950s supermarket product aisle, and those shelves of child-rearing guides at the bookstore are today's equivalent of *Good Housekeeping*, offering women the possibility of earning a doctorate in mothering.

The rebellious reactions to these different standards are tailored to their eras. In the late 1960s and '70s, women rose up against perfect housewifery. Sue Kaufman wrote *Diary of a Mad Housewife* in 1967; in 1973, Erica Jong wrote *Fear of Flying*, which included a rant about the ideal woman: "She cooks, keeps house, runs the store, keeps the books, listens to everyone's problems. . . . I was not a good woman. I had too many other things to do." Today, on the other hand, the typical rebellion story is not about being a bad wife. It's about being a bad *mother*—which in fact is the title of a 2009 book of essays by Ayelet Waldman.

Tales of maternal malfeasance capture our imaginations because the imperatives of "intensive mothering" still persist, driving mothers to seek moral reassurance in all sorts of ways. Hays, for instance, notices that whenever stay-at-home mothers confess to ambivalence about the choice they've made, they justify their decision by saying that staying home is best for their kids. But whenever working mothers confess to ambivalence about the choice they've made, they say the exact same thing: *working* is best for the kids. "The vast majority of these women do not respond," Hays writes, "by arguing that kids are a pain in the neck and that paid work is more enjoyable." Instead, she says, they argue that they are providing their children with added income for extracurricular pursuits. Or that they are modeling a work ethic. Or that work makes them more focused parents and improves the quality of the time they spend with their kids. They use the effect on the child, every time, to justify any answer.

the irreproachable single mom

"She might or might not wear this," says Cindy Ivanhoe, pulling a dress from the rack at JC Penney. "Sar? Sarah? Sarita?"

Her daughter, arty and adorable, looks up from her own browsing. She wrinkles her nose.

"I can Botox that face for you," says Cindy. She's a doctor, which makes this a pretty good joke, though her specialty is brain injuries, not plastic surgery. For a long time, Cindy worked at Houston's TIRR Memorial Hermann, helping to design the program in which Congresswoman Gabrielle Giffords ultimately enrolled to rehabilitate after sustaining grave injuries in a mass shooting in 2011. A few years ago, Cindy transitioned to private practice, though she still writes and teaches at Baylor College of Medicine.

Sarah picks up a dress that's black and white and somewhat provocative. "Uh-uh," says Mom. "You're *twelve.*" Her daughter needs something for a friend's bat mitzvah.

Cindy turns to me. "I think the fatigue is the hardest part," she says. "Because you don't get to be the person you want to be. It'd be useful if I didn't work, but on the other hand, I'd be crazed if I didn't work. It's become my break, which is pretty sick—"

Sarah shows her something that looks like a skirt, though it's hard to tell for sure.

Cindy stares. "No. Those look like bloomers."

"I have the shirt for it."

"You have the shirt. But . . . no." She sighs.

Who would she be, I ask, if she weren't tired?

"I'd be less grumpy by the end of the night, when I have to check homework," she says. "Or listen to her chant for her bat mitzvah, or try to get a teenage boy"—her son, she means—"to talk a little bit about what he's really feeling, because he suppresses his stress. I'd have more time for it. But I have all the responsibilities." Cindy and her husband separated in 2006. He no longer lives in Houston.

Cindy picks up a long, flowing, sixties-style sundress. "I promise if you like something like this, I'll get it shortened."

Sarah responds by holding up an abomination in tulle. "I'd honestly rather have *this* than *that.*" She's smiling, not being insolent.

Cindy nods. "Agreed." We head into the dressing-rooms and take a couple of seats in the common area. Sarah disappears behind one of the curtains. "It took me years to turn off work when I got home," Cindy continues. "Partly, it was having kids that gave me that. But because of other stressors—financial and whatever—I'm still . . ." She doesn't finish the sentence, but the word is implied: *tired*.

Sarah starts to yo-yo in and out of the dressing room, modeling dresses. ("No." . . . "No." . . . "Maybe." . . . "Cute." . . . "Too short." . . . "You'll outgrow it in weeks.")

"And I kind of worry," says Cindy, "that their memory of me—what's it going to be? 'Mom was always trying to get her work under control. Mom was always trying to get the bills paid on time.' Or whatever. 'Mom was always trying to get it better.' "

Then out pops Sarah, this time without a hint of irony on her face. It's obvious why: she looks terrific. Funnier still, she had grabbed this dress as a joke. It's "froufrou," as she likes to say, a traditional satin number. She'd taken it to amuse herself. Now she's dazzling us.

Cindy beams. "Let me fix your straps." She fiddles with them. "Do you feel pretty?"

Sarah nods, still shocked.

"Turn around." Cindy admires her. "Do you want it for this Saturday night's party?"

"Yes."

Sarah spins on her heel and struts back into the dressing room. Cindy's eyes follow her. "You know, you bring kids into the world. You love them. You'd *die* for them." She shakes her head. "I just wish I got more rest."

STUDIES ABOUT HAPPINESS AND parenting are tremendously complex, and often problematic. But no matter how they're done—no matter what methodology researchers use, or what data set, or what piece of the heart or soul they attempt to translate into numbers—single mothers do not do well.

The reason for this, most likely, is economic. Single mothers include not just divorcees but women who have never married. As a rule, mothers who have never married have also never gone to college and therefore have dramatically foreshortened economic horizons: they have less than one-fourth the income of two-parent families, they have more health problems, and they have fewer social ties. Given these brutal facts, it stands to reason that never-married women would affect the results of any survey that attempts to gauge maternal happiness.

Cindy's case is different. She is divorced, not never-married. She earns a good living, lives in a fine house, and has a loyal circle of friends. Yet her life, which is filled with middle-class comforts, shows why divorced mothers, too, may still suffer more than their married peers. It's true that they're more likely to receive child support. But their income, on average, is still only half that of two-parent families. (Cindy is also wrestling with some tricky finances right now, having signed a lease on a new medical office just before the crash of the real estate market.) Because the children of divorce tend to spend more time under their mothers' roofs than their fathers', single mothers bear the brunt of the emotional work that comes with easing kids through the separation—a special variation on what Arlie Hochschild calls the "third shift," in which mothers are the designated family empaths. And middle-class single mothers are burdened by the same intensive parenting pressures as married mothers, but with less time and flexibility to manage them. Suzanne Bianchi, who has so agilely crunched numbers from the American Time Use Survey, notes in a recent paper that single parents often "have as many demands on their time as married parents but half as many adults to meet those demands."

Changing Rhythms of American Family Life, a 2006 compendium of data about family time-use co-authored by Bianchi, tells this story in numbers. Single mothers are more apt to report having too little time for themselves than their married peers (especially married fathers), more apt to report multitasking "most of the time." They spend four and a half fewer hours each week on socializing than their married counterparts, and one and a half fewer hours having meals. "Sometimes I'll meet a friend for a drink," Cindy tells me. "Verrrrry rarely I'll get to a movie. But I'm always behind."

These issues of time, which translate so quickly into issues of stress, spill over into other areas of life. Like dating. "I had gone out with someone I met this summer," Cindy tells me as we're sitting in the dressing room, "and I said to him, 'You know, by the time school starts, you're going to run from the burning building. Come September or the end of August, whatever freedom I have right now goes *poof*.' "

A few minutes later, she gets a text from her son, whose knees have been bothering him—he runs cross-country at school. *Can you pick up a ton of ice on your way home?* She looks up from her phone and stares into space. "Like I've told many a friend, I wouldn't date me," she says. "It's like, 'Sorry, I have to go get ice now.' " As soon as we leave JC Penney, she runs off to go get it. Who else is there? Then she goes home and carefully wraps her son's knees.

a man's work is never done

It's not just women experiencing undue pressure to be perfect mothers ("the mommy mystique," as Judith Warner calls it in *Perfect Madness*). Men are experiencing these pressures too. The Families and Work Institute calls this phenomenon "the new male mystique" in a report by the same name issued in 2011. Based on a large, nationally representative sample of the American workforce, the organization found that today's fathers work longer hours than their counterparts without kids (forty-seven hours per week versus forty-four) and that they're far more likely than non-dads to do fifty-plus-hour workweeks (42 versus 33 percent). Most surprising, however, is the report's finding about work-family conflict: today men are more apt to experience it than women, especially if they're in dual-earner couples.

Part of the reason men feel this way can be explained, not surprisingly, by the uncertainties, rigors, and excesses of the modern economy. The men in this study fretted much more about losing their jobs than they had in previous studies by the Families and Work Institute. In 1977, 84 percent of male respondents believed themselves to be professionally secure; in 2008, even before the recession, that number was down to 70 percent. Today's workers also have to withstand ritualistic work intrusions into their homes at all hours via technology: 41 percent of the sample reported getting office messages during non-office hours at least once a week. And in 2008, far more men agreed with the statement "At my job I have to work very hard" than in 1977 (88 percent versus 65 percent). More also agreed with the statement "At my job I have to work very fast" (73 percent versus 52 percent).

But Ellen Galinsky, a co-author of the study and head of the Families and Work Institute, suspects that fathers today are also experiencing a shift in cultural priorities, and with it, a shift in internal priorities. "They don't want to be stick figures in their children's lives," she told me.

No one showed this more vividly than Steve Brown as he sat there on the soccer field, watching his son. His phone was a pinging, buzzing pinball machine, and he was scrolling through emails throughout the practice. ("This is usually BlackBerry time," he apologized.) At some point, I asked him what he found hardest about being a parent. "Finding the time to do everything you want to do," he immediately answered. "Work-life balance. And even community balance now. Sometimes all three of those things." Community, because Steve could be attending fundraisers and political events five nights a week, if he chose. "And it's really a challenge for us to figure out," he says. "Like, which weeks can I be Mr. Chairman? What days do I need to come home? And what days does Monique need to go somewhere? I have to pick and choose."

And Steve has more job flexibility than most men. As the head of his own shop, he sets his own hours; he's able to work on weekends if he can't compress the load into the normal workweek. But the thing is, Steve doesn't *want* to work weekends. He wants to be on the soccer field, because it's game season and his older son is a good soccer player. But that means refraining from pursuing larger political aspirations—which he has—for the sake of family stability. "Now's not a good time to go to DC or Austin," he told me. "At some point, we'll make that decision, but when these guys are a little older."

A previous generation of men never thought this way. Some men still don't. And for all his progressive instincts, Steve readily concedes that Monique still does most of the work around the house. She does most of the child care too, though one of his favorite jokes is to make the deadpan declaration, "I'm the primary caregiver," just to see how Monique will react. She cooks, though he does the dishes. She tends to bathe the boys and prepare their outfits for the following day, though occasionally it's the other way around. Later that evening, when I speak to Monique, a social worker in downtown Houston, she tells me the exact same thing that Kenya, one of the ECFE mothers, told me: "The most stressful part of my day is from five, when I leave work, until ten."

This past year, the boys spent two weeks at her parents' in Baton Rouge, going to tennis camp. "And we worked late every day instead of going out," she says. "We were catching up on meetings we never had and errands we never did."

I ask how her professional life has changed since having children. "I used to work with foster kids," she says. "That required a lot of evening work. I could never do that now. But I *loved* working with them."

She looks at her children. It's getting to be that hour when kids start flapping blindly about the house like moths. "Sometimes I wish I could put glue on their butts." She starts laughing. "Managing the two of them is four times the work of managing one."

"I disagree," says Steve, who's discreetly eating takeout and watching the US Open in a corner of the kitchen. "I say two."

Monique rolls her eyes as if to say: *This guy has no idea.*

Steve looks her dead in the eye. "I'm the primary caregiver."

Monique returns the look. "If you're the primary caregiver, did you know that Mathis"—their three-year-old—"came in at two-thirty in the morning, wet?"

His eyes widen for a brief second. "No."

"He'd had an accident. I got up, changed him, and changed the sheets."

Steve had slept through it. And she didn't hold it against him. They had a thing going, these two, an arrangement that worked. "Primary caregiver," she says, smiling at him.

the indoor child

There aren't a lot of children riding bikes around here. It has taken me a few days in the Houston suburbs to diagnose this strange absence, this latter-day version of the dog that doesn't bark. Today is a sultry and sunshiny day, the kind that during my own childhood would have been cause for my mother to shoo me into the cul-de-sac for the rest of the afternoon. But here on this lushly landscaped street, just footsteps from Palmer Elementary School in Missouri City, Texas, all is quiet.

I think about this as I approach the door of Carol Reed's house, a pretty brick structure with a small pool in the backyard. There's just so very little street life.

Carol greets me at the door. She, like Monique, is on the Palmer

PTO, though she lives in a slightly tonier neighborhood, with slightly pricier homes. But unlike most of the women I've met in this area, she grew up in Massachusetts (with the accent to prove it) and prefers her hair short and her glasses chunky. And unlike most women I've met here—or anywhere—she's raising children a full generation apart from one another. She had her son at just twenty-one. Six years ago, when she was forty-seven, she and her second husband decided they wanted a child together, and they adopted a baby girl from China. When Emily first arrived, she was malnourished and had just had heart surgery. Today she's a hale first-grader. Carol stays home with her.

Carol gives a mixed report about the differences between parenting then and now. She has a bigger social network now, which is nice, and more confidence and experience—others don't try to tell her how to do her job. "But because Emily's an only child," says Carol, "she wants me to be her playmate."

Yet her son, now thirty-one, was also an only child, I point out. Why was it so different? "I don't know," she says, after a moment. "But I played with him less. There were more neighborhood kids. More sleepovers." She thinks about it some more. "Emily likes having people here. But not going over to other people's houses."

It's possible, of course, that Emily is that kind of kid ("Mom! Play with me!"). But as Carol gives me a tour of her home, I develop another theory. Emily doesn't just have her own beautiful bedroom. She has her own beautiful playroom too, dominated by a yellow dollhouse and a giant easel and a nifty kitchen. Art supplies, stuffed animals, and toys fill every corner. They're stacked in translucent drawers and brightly colored crates; they're tucked inside a banquette that runs along the window; they're piled high on all the play surfaces, including her miniature table and chairs. The place is a child's wonderland, indistinguishable from many preschool play spaces and pediatricians' waiting rooms. The strange thing is that this room is not all that different from the playrooms of many other middle-class children. It's no longer exotic, having all this equipment, all these toys. They're manufactured abroad and sold here at affordable prices, either through Amazon or at the local Walmart.

"This is like her apartment," says Carol as she walks me through it. "It's where she entertains us. She changes it every now and then. Sometimes it'll be a restaurant and she'll make you coffee or a smoothie or cakes"—she points to the toy coffeemaker, blender, and mixer—"and sometimes it's a store." She points to the toy shopping cart and cash register.

I look at Carol.

She laughs. "I know," she says, looking at everything. "My son didn't have any of this."

THE SENTIMENTALIZATION OF CHILDHOOD has produced a great many paradoxes. The most curious, however, may be that children have acquired more and more stuff the more useless they have become. Until the late nineteenth century, when kids were still making vital contributions to the family economy, they didn't have toys as we know them. They played with found and household objects (sticks, pots, brooms). In his book *Children at Play,* the scholar Howard Chudacoff writes, "Some historians even maintain that before the modern era, the most common form of children's play occurred not with toys but with other children—siblings, cousins, and peers."

But by 1931, kids had enough gear for the Hoover White House to declare that they deserved a room of their own. Children, said conferees on a panel on child's health, needed "a place where they may play or work without interference from or conflict with the activities of the adult members of the family." The idea of the modern playroom was born, by executive decree.

In the years directly following World War II—the time when modern childhood began in earnest—the toy boom began in earnest too. In 1940, toy sales were a modest $84 million; by 1960, they had reached $1.25 billion. Many classic children's toys were invented during this era, including Silly Putty (1950) and Mr. Potato Head (1952). And the pickings back then were paltry compared to today, when playrooms as well stocked as Emily's are increasingly common. In *Parenting, Inc.* (2008), Pamela Paul writes that toy industry sales "for babies between birth and age two alone" were over $700 million annually. According to the Toy Industry Association, domestic sales of kids' toys were $21.2 billion in 2011, a figure that didn't include video games.

Such oceans of plenty have had unintended consequences. In *Huck's Raft*, Steven Mintz notes that toys before the twentieth century were primarily social in nature—jump ropes, marbles, kites, balls. "Modern manufactured toys," on the other hand, "implied a solitariness that was not a part of childhood before the twentieth century." He's thinking of Crayons, for instance, introduced in 1903. Or Tinker Toys (1914), Lincoln Logs (1916), or Legos (1932).

More generally, writes Mintz, "one defining feature of young people's lives today is that they spend more time alone than their predecessors." They grow up in smaller families (22 percent of American children today are only children). They are more likely to have their own rooms than children in generations past, and to live in larger homes, which means the very architecture of their lives conspires against socializing with other family members. They also live in a nation of suburbs and exurbs, where neighbors and friends live farther away.

Isolation results in a lot of extra work for parents. Their children recruit them as playmates, as Emily does Carol. They are prodded for rides hither and yon. And parents oblige, worrying that their children will suffer from loneliness if they don't. This is yet another reason why mothers and fathers schedule so many after-school activities for their children. Lareau noticed it immediately in the families she studied. "Middle-class parents," she writes, "worry that if their children do not enroll in organized activities, they will have no one to play with after school and/or during spring and summer breaks."

The result, unfortunately—and entirely inadvertently—can be a self-perpetuating cycle. If kids lead tightly scheduled lives from the time they're young (including preschool, which increasingly takes a modular approach to dividing the day), they seldom experience boredom, which means they don't really know how to *tolerate* boredom, which means they look to their parents to help them alleviate it. Nancy Darling, an Oberlin psychologist and author of the sterling parenting blog *Thinking About Kids,* made this point in a 2011 post. When she was a child, she notes,

> we were bored all the time. There were no extracurricular activities for kids until junior high except for Scouts once a week or maybe 4H and Sunday School. Few moms worked, so we came home from school at 3:00 and just hung out. They hadn't invented Sesame Street yet and Bugs Bunny and Rocky & Bullwinkle were more or less all of kids' television unless it was Saturday morning. . . . What that meant is that our moms—who were busy cooking, cleaning, watching soap operas, hanging out with their neighbors, and generally running a huge network of non-profit services (Scouts, Church, Red Cross, etc. etc.) would typically respond to our complaints that we had nothing to do by suggesting that our rooms could definitely use cleaning. We learned not to ask and figured something out.

So it's more than a little maddening when our own children can't seem to do the same, even if we've played a role in diminishing their capacity for resourcefulness. It's not that these organized activities don't have their virtues, adds Darling. But because of them, she speculates, "kids have very little experience learning to find things to do FOR THEMSELVES. They have been PASSIVE [capital letters hers, not mine]." This passivity can be especially hard as kids exit the elementary school years, and the burden of figuring out how to direct their free time becomes their responsibility. As Darling later explained to me: "No kid ever says, 'Wow, now I've got some free time on my hands—I'm going to be a stamp collector!' Hobbies take time to develop."

But because middle-class children today occupy privileged positions within the family, and because their parents have overextended themselves on their behalf, kids sense that they have the power to make their boredom their parents' responsibility. Lareau noticed this immediately too. "Middle-class children," she writes, "often feel *entitled* to adult attention and intervention in their play."

PARENTS WOULD DOUBTLESS FEEL a lot less pressure to keep their children busy or entertained—and more confident about their kids' ability to make their own fun—if they felt comfortable sending their children outdoors. But increasingly they don't. Here, again, is another paradoxical consequence of our sentimentality: the more economically useless children have become, the more aggressively we've tried to protect them.

One can discern the outlines of this trend by simply studying the history of the modern playground. In 1905, there were fewer than 100 playgrounds nationwide. By 1917, there were nearly 4,000, because reformers had agitated mightily for them. Before then, children played in the streets. But suddenly they needed protection from a brand-new and lethal invention: the automobile. One of the solutions was the Playground Association of America, established by reformers in 1906.

Today, children lead even more cloistered lives. They grow up in homes with padded coffee tables, plugged-up electrical sockets, and gated stairways. They go to playgrounds that offer protection not only from the streets but from their own equipment, with swings as snug as diapers and spongy surfaces to break falls from the jungle gym.

So perhaps it's no surprise that by the time children get big enough to venture out on their own—to the grocery store, to a friend's house down the street—their parents feel strange about letting them go, believing the world to be a dangerous place. The number of elementary- and middle-school students who walk or bike to school dropped from 48 percent in 1969 to 13 percent in 2009, according to figures from the Department of Transportation, even though crimes against children have been steadily declining for the last couple of decades, making this moment in time perhaps as safe for children as it's ever been. (To name one example, between 1992 and 2011, reports of child sexual abuse fell by 63 percent.)

It's also possible that this anxiety about child safety is yet another manifestation of our culture's ambivalence toward women in the workplace. With so many mothers collecting a wage outside the home, there are fewer eyes on the street, and with fewer eyes on the street, a panic blossoms about the potential dangers of those streets. Mintz notes that throughout the eighties—the decade when women were marching off to work in their Reeboks and power suits—paranoia ran rampant about sexual abuse in day care centers. "In retrospect," he writes, "one can see how terrified parents displaced their own anxieties and guilt feelings about leaving their children with strangers onto daycare workers." There was a near-simultaneous wave of alarm over stranger abductions and madmen inserting razor blades into Halloween candy; the faces of missing children began popping up on milk cartons at around this same time too, when in fact the number of abductions by strangers probably numbered between 500 and 600 per year, or one in roughly 115,000 (while approximately four times as many kids died as passengers in car accidents).

Today, abduction paranoia is stoked not so much by milk cartons as the excesses of cable news and a new transparency in criminal records. Both forces were especially evident in Texas. Nearly every parent I spoke to in Sugar Land and Missouri City mentioned their kidnapping fears to me at some point or another, even though they lived in ravishingly secure middle- and upper-middle-class neighborhoods. I soon learned that Texas has a public, Internet-searchable sex-offender registry. Anyone can go online, type in his or her address, and see where the nearest recently released sex offender lives. Carol Reed speculated that it was one of the reasons her older child needed less attention when he was growing up. "Things weren't as scary as they are now," she told me. "Or we weren't *aware* of how scary it was."

I'll confess I found this fear intensely irrational at first, given the robust health of her neighborhood and her home's proximity to the local school. But then I checked online. An address and profile of a fellow just a half-mile from her house immediately appeared, followed by three others, all under one mile away. The details of their criminal histories were not very specific. Roughly 90 percent of convicted sex offenders have assaulted people they know, rather than the children of strangers. But this is hardly the type of news that reassures most parents. I'm not at all sure I find it reassuring myself.

SAFETY CONCERNS MAY BE partially responsible for keeping children indoors. But so does an emperor's gallery of electronic entertainments, which almost no modern family lives without.

"Half the time," a Sugar Land mom told me, "if my son"—he's ten—"goes outside, he'll come back just a few minutes later with the neighbor boy to play video games."

The technological developments of the last fifteen years are obviously huge, and I'll be exploring them in more depth in the next chapter. I should also declare from the start that I'm not an alarmist when it comes to the new modalities of the information age. But if one simply considers the raw data about video games (a 2010 Kaiser Family Foundation study says that eight- to ten-year-olds play them for roughly an hour each day) and then adds the data about television for the same age group (3 hours and 41 minutes, up 38 minutes since 2004) and non-school-related computer use (46 minutes), one can see why parents would start to wonder about the cumulative impact of these entertainments—especially for a generation that has a hard time tolerating boredom. (Sixty percent of all "heavy" media users in the Kaiser study described themselves as "often bored," compared to 48 percent of "light" users.) Screen time consumes a lot of time. Some of it is social, but more of it is solitary. A recent study found that 63 percent of seventh- and eighth-grade boys "often" or "always" play video games alone.

Fears about this new and intense form of time-use provide yet another explanation for why parents sign up their children for more familiar organized activities. And it again explains, I think, the huge appeal of scouting in many of the communities I visited.

The Boy Scouts of America were founded in 1910, a time of rapid urbanization, which prompted fears that young men were degenerating into dandies, opting for the easy pleasures of city life over the rough work of the farm. It was also at just this moment that child labor was being criminalized, and children were, as Viviana Zelizer ruthlessly puts it, becoming useless.

The result was an almost hysterical fear of male softness. And this fear still exists today, often expressed when parents talk about their kids' excessive Xbox-playing or Hulu-watching. The Boy Scouts seem the perfect antidote to those sedentary hours spent in front of a screen. "I *love* Scouts," said Laura Anne, the Scout mom from the beginning of this chapter. "We did Scout camp last week. Andrew and Robert got to do the things boys don't do anymore. Leatherwork, archery, firing BB guns—old-timey fun."

It doesn't matter that gaming and online adventures may turn out to be *useful* for the next generation, preparing them for a future written in HTML code. On some primitive level, rightly or wrongly, we still associate practical skills with things you can *physically do.* It's nostalgia for these more tactile, "manual" enterprises that surely explains the phenomenal popularity of *The Dangerous Book for Boys,* published in 2006. It gave instructions for how to *do* things: How to tie five essential knots. How to hunt and cook a rabbit. How to make a go-cart, a battery, a treehouse. To adults, it's much harder to see the value of video games.

But a child can. From a child's point of view, video games provide great opportunities for flow. They provide structure and rules. They offer feedback, telling players how well they've done. Video games supply a chance to excel at something, to gain a sense of mastery. "There's now this weird structural tension," says Mimi Ito, a cultural anthropologist at the University of California at Irvine who studies technology use. "We've seen a heightened arms race to good educational pathways and good jobs. So kids look to these online spaces for the autonomy they've lost"—going to all these structured activities, she means, in order to win that arms race—"while parents are more focused on efficiency and look at these spaces as completely time-wasting."

And she wonders too if kids would be less inclined to sink hours into these indoor entertainments if they were given more freedom outside the home.

the burden of happiness

Angelique Bartholomew, forty-one, lives just a few houses down from Carol Reed, in a similar brick structure, also just a couple of blocks from Palmer Elementary School, where she too serves on the PTO board. The challenges inside her home, however, are different. Carol struggles with the intense dynamic of keeping one child entertained; Angelique has four children under her roof, plus a stepdaughter who frequently stays over, which means her role is often indistinguishable from that of an air traffic controller. This afternoon, it's fairly quiet. Her thirteen-year-old, Myles, is at football practice. Her nine-year-old, Brazil, is at a piano lesson. Her youngest, Niguel, is napping. So Angelique, a stunning African American woman in giant hoop earrings and bare feet, is taking advantage of this atypical hour of calm to prepare dinner. On a more ambitious night, it takes two chickens, seven yams, and a crate of strawberries to feed her crew, but this evening she's keeping it simple with tacos. As she stirs a pan of ground turkey, her four-year-old daughter, Rhyan, wanders downstairs and starts vigorously jumping up and down on the living room couch. Angelique looks up at her.

"Should we get you a book or crayons, and you can do some coloring?"

"My book's up there." She points.

"*Please.* Manners. *Please.*" Angelique fetches the book. "We're having an adult conversation. Don't butt in."

I ask Angelique what she finds hardest about having a large family. I expect her to say managing everyone's schedule, or making her marriage work, or surviving her mortgage payments, or getting enough sleep, or carving out enough time for herself, or keeping her career alive. (She works part-time for a medical forensics lab, but her true love is inspirational speaking.) None of these is the answer I get.

"Balance between the kids," she instantly replies. "Making sure they all feel like they're important. Because I know which one of them doesn't feel like they're as important."

She's speaking candidly while still treading cautiously, daring to say something many parents won't, but carefully concealing the identity of the child by using the plural pronoun "they." "All my kids are self-sufficient," she continues. "But this one wants me to slow down and take a minute and acknowledge them. And I'm in continual motion. . . ."

Even now, in this hour of quiet, a pile of paperwork for a big meeting she has tomorrow needs attention, as does a pile of PE forms for her oldest. Her husband, who sells medical equipment to hospitals, is in San Antonio, which means he can't help out, and her sister—Angelique is one of ten kids—on whom she heavily relies, is asleep upstairs because she works nights.

Yet Angelique and her husband have managed to make it work. They've got a large, airy home and a beautiful pool out back with a twisty slide. They can afford nursery school. She finds time to do her work during the day; she volunteers for the PTO and a local organization that supplies clothes to schools. And she keeps her sanity by waking each morning at six o'clock to meditate and pray.

It sounds tiring—*physically* tiring. But what's most exhausting to her, she says, are the emotional demands of raising children. "It's funny how the same parents can make totally different characters," she says. "No matter how much you put into some of your kids, some will need more than others. And this one"—the one who's on her mind at the moment, the one to whom she obliquely referred before—"will always require more." She starts chopping strawberries for dessert and shifts the discussion to the more general theme. "I want them *all* to know that they're really important to me," she says. "I feel the same love and care and nurturing for each of my kids. It's just that the relationship with each is different."

What does she do to make each of them feel important? "I'll ask one to come to the store with me," she says. "Or I'll act like I can't do something, and ask one to help me. Or I'll lay in the bed with one. That's a big deal. Saying prayers before going to sleep." She throws open the fridge, inspects it, frowns. Not enough cheese. "And then when *I* go to sleep," she continues, "I'll think of something I said to one, or my response to another, or my reaction to another . . . and I'll try to get up the next day and go to that kid first, if I don't like what it was."

I don't know why I ask her this next question, exactly, but it feels natural. What, in her view, makes a good mother?

She stops what she's doing to consider this riddle. "Looking at my kids," she says after a while, "and saying, 'Are you hungry? Are you sad?' Identifying the emotion." She resumes her work, pulling out a box of eighteen taco shells and thunking it down on the counter. "Identifying the emotion," she repeats for emphasis. "The mother who can detect before it's spoken. See the place the kid's in before the kid says it. That's a good mom to me."

IN MANY WAYS, ONE expects a modern and conscientious mother to be like Angelique. She should not play favorites. She should be mindful of her children's sensitivities. And above all, she should make her children feel important, building their self-esteem shingle by shingle, block by block.

But "modern" is the key word here. Before the "sacralization" of childhood (another apt description from Viviana Zelizer), parents' hearts weren't expected to double as emotional seismographs. It was enough that they mended their kids' clothes, fed them, taught them to do good, and prepared them for the rigors of the world.

It was only after parents' primary obligations to their kids had been completely outsourced—to public schools, to pediatricians, to supermarkets, to the Gap—that the emotional needs of their children came sharply into focus. In *Raising America,* Ann Hulbert cites the 1930s sociologist Ernest Groves, who observed: "Relieved of having to carry out all the details of child-care in all their ramifications, the family today can concentrate on the more important responsibilities which no other institution can perform—direction, stimulation, and loving friendship."

But what, exactly, does it mean to provide "direction, stimulation, and loving friendship"? These are, to say the least, abstract objectives. Yet practically every child-rearing expert has insisted on them since World War II. "Stimulation and loving friendship" was the central lesson of *Mary Poppins* almost half a century ago—the character of George Banks metamorphoses from a distant Edwardian paterfamilias to an emotionally engaged maker of kites (a lesson almost every movie dad has learned since)—and it is the central tenet of almost all parenting blogs today. (For years the capsule description of the *New York Times* parenting blog Motherlode started like this: "The goal of parenting is simple—to raise happy, healthy, well-adjusted kids.") In *The Cultural Contradictions of Motherhood,* the sociologist Sharon Hays sums up her close reading of the works of Benjamin Spock, T. Berry Brazelton, and Penelope Leach, three of the most popular child-rearing experts in history: "Individual happiness becomes that elusive good upon which we can all agree."

I should here point out that individual happiness is precisely the goal I have for my own son too. But in one of his essays, Adam Phillips, the British psychoanalyst, makes an observation I've never quite been able to shake:

> It is unrealistic, I think—and by "unrealistic" I mean it is a demand that cannot be met—to assume that if all goes well in a child's life, he or she will be happy. Not because life is the kind of thing that doesn't make you happy; but because happiness is not something one can ask of a child. Children, I think, suffer—in a way that adults don't always realize—under the pressure their parents put on them to be happy, which is the pressure not to make their parents unhappy, or more unhappy than they already are.

Parents probably wouldn't be so frantic about making children happy if their children had more concrete roles within the family. Writing in 1977, Jerome Kagan remarked that the modern, useless child cannot "point to a plowed field or a full woodpile as a sign of his utility." Hence, he predicted (with uncanny prescience), children were at risk of becoming overly dependent on praise and repeated declarations of love to build their confidence.

Nor would parents be so anxious about shoring up their children's self-esteem if, to use Margaret Mead's phrase, "the sureness of folkways" guided their efforts and they knew what, precisely, they were preparing their children *for.* Dr. Spock, the first child-rearing expert to write in the era of the protected child, discusses this predicament in *Problems of Parents* (1962), and it's probably no accident: he was the pediatrician for Mead's daughter. "We are uncertain about how we want our children to behave," he writes, "because we are vague about our ultimate aims for them." Unless we've had "unusually purposeful" upbringings, he says, American middle-class parents

> fall back on such general aims as happiness or good adjustment or success. These sound all right as far as they go, but they are quite intangible. There's little in them that suggests how they are to be accomplished. The trouble with happiness is that it can't be sought directly. It is only a precious by-product of other worthwhile activities.

This, I think, explains why Amy Chua's *Battle Hymn of the Tiger Mother* was such a titanic success. It preached exactly this same idea. Forget all this airy-fairy talk about happiness. Aim for excellence instead. Happiness from a job well done is the best kind of happiness anyway. It leads to lasting esteem.

The irony is that even Chua has questions about this approach. "If

I could push a magic button and choose happiness or success for my children," she writes on her website, "I'd choose happiness in a second."

homework is the new dinner

"That's a good shovel—where'd you find that?" asks Laura Anne. Cub Scout sign-ups are over, the kids have eaten dinner, and now we're all sitting at the kitchen table: it's homework time. Robert, Laura Anne's seven-year-old, is working quietly on more modest assignments. But Andrew, her nine-year-old, has to transform a big cardboard doll into some kind of scientist. He's opted for an archaeologist. He nimbly tweezes the shovel between his fingers and affixes it to the doll. "From my Lego set," he answers.

"And what do you want me to work on?" asks Laura Anne. "What else does he need?"

Andrew draws a beard on the doll, plus some shorts and a belt. "Look!" He adds a gray hat.

"I love that! Where are your oil pastels? He needs some dirt." She leaps up and pulls them out of a cabinet.

At this point, I ask Laura Anne: Why is she offering to help with this? With all due respect to Andrew, this project is, in the end, just a doll.

Laura Anne says she's aware of that. But it's her habit now. Some of the projects the teachers assign are much more involved, practically demanding parental assistance; she'd feel remiss if she just sat around while he did them. "Let me show you the Scotland project," she tells me and heads out to the garage. This is an ancestry project that all the kids had to do the previous year. She returns a few minutes later with a spectacular black triptych. "Scotland, by Andrew Day" it says on top. A kilt hangs like a pageant sash over its center panel. "I couldn't throw it away."

It's certainly impressive, dappled with photos and essays under tidy headings: "Land and People." "An Interview with a Modern Day Bagpiper." I ask if it was the snazziest-looking assignment submitted that day.

She shakes her head. "There was one with a giraffe that went all the way up to here." She stretches her arm as high as it'll go. "And there was this other project where the kids had to do a building in the city, and one boy did the retractable roof" (of Reliant Stadium). She runs the kilt through her hand. "I was *making* a kilt in case I couldn't get one," she says. "I had the sewing machine out with the pattern. . . ."

Robert interjects at this point that he's finished his assignments. Laura Anne looks them over. "Wow. Those are good sentences. You finished your homework packet, and it's only Tuesday!"

She takes a seat and resumes working on Andrew's archaeologist doll. The table has been given over completely to art supplies, notebooks, workbooks, markers, and pencils. It could be a workbench in a classroom. "I think homework has replaced the family dinner," says Laura Anne. She lets the observation hang for a moment, then adjusts the archaeologist's shirt. "Maybe it's sad, but it's true. Because this is when your children tell you stuff. This is the time you're sitting down with your children and creating something with them." She admits that one of the reasons homework may have replaced dinner in her family is that she's not all that big on cooking. In a city, it's easy not to be. The kids ate takeout tonight, and the Styrofoam detritus is still scattered about. "I always knew my mom cared about me because she fed me, right?" She looks up from her son's project. "She put love and time into the meal. But I'm not like that." Housewifery was for her mother's generation. Her generation transforms their kitchens into homework outposts. She snips a strip of fabric and hands it to her son. "So this is me," she says, "doing my gifts of service. Putting in love and time."

SUZUKI, A METHOD OF musical instruction first developed in the aftermath of World War II, was designed to teach very young children how to play the violin. At the heart of the method lay a very generous theory, which is that all children are capable of musical accomplishment if given the right tools, techniques, and environment. The Suzuki method requires a high level of commitment from a child. But what makes it truly unusual is that it requires a high level of commitment from parents too. Parents must attend music lessons and pay attention to what's being taught. They must supervise practice every day. They must immerse their children in a musical environment, playing symphonies in the home and taking them to concerts during their free time.

Today, people use the Suzuki method to teach all kinds of musical instruments, not just the violin. It also serves as a pretty good metaphor for the way middle-class parents approach their children's activities. Everything they do has to be full-saturation involvement, done side by side. It's not just violin. It's making the pinewood derby car for Cub Scouts. It's playing the role of sports agent during travel-team season. It's curating summers at six different kinds of camps. It's playing restaurant when your kid's bored. It's doing Kumon drills. It's working as collaborators on school projects. It's signing off on assignments, a practice more and more schools seem to require. *Homework is the new family dinner.*

But what gets lost in all this?

One wonders if *actual* family dinners, whose numbers have fallen quite a bit since the late seventies, might happen a bit more frequently if they hadn't been supplanted by study halls at the dining room table, and if that time wouldn't be more restorative and better spent—the stuff of customs and stories and affectionate memories, the stuff that binds.

Family time isn't the only thing that suffers under this new arrangement. Couple-time suffers too. If homework is the new family dinner, soccer practice is the new date night (itself a modern invention).

Steve Brown said as much, as we were sitting on the sidelines, watching his son play. "Last week, my wife and I were able to do this together," he told me. "This, uninterrupted, is good mommy-daddy time to have a conversation." A breeze swept through, fluttering the trees and rippling the grass. He shut his eyes.

And I did see what he meant. The sun was setting, the shade was cool, his handsome kid was playing a glorious game. But there has to be a better venue for mommy-and-daddy time than a soccer game. Parenting pressures have resculpted our priorities so dramatically that we simply forget. In 1975 couples spent, on average, 12.4 hours alone together per week. By 2000 they spent only nine. What happens, as this number shrinks, is that our expectations shrink with it. Couple-time becomes stolen time, snatched in the interstices or piggybacked onto other pursuits.

Homework is the new family dinner. I was struck by Laura Anne's language as she described this new reality. She said the evening ritual of guiding her sons through their assignments was her "gift of service." No doubt it is. But this particular form of service is directed inside the home, rather than toward the community and for the commonweal, and those kinds of volunteer efforts and public involvements have also steadily declined over the last few decades, at least in terms of the number of hours of sweat equity we put into them. Our gifts of service are now more likely to be for the sake of our kids. And so our world becomes smaller, and the internal pressure we feel to parent well, whatever that may mean, only increases: how one raises a child, as Jerome Kagan notes, is now one of the few remaining ways in public life that we can prove our moral worth. In other cultures and in other eras, this could be done by caring for one's elders, participating in social movements, providing civic leadership, and volunteering. Now, in the United States, child-rearing has largely taken their place. Parenting books have become, literally, our bibles.

It's understandable why parents go to such elaborate lengths on behalf of their children. But here's something to think about: while Annette Lareau's *Unequal Childhoods* makes it clear that middle-class children enjoy far greater success in the world, what the book can't say is whether concerted cultivation *causes* that success or whether middle-class children would do just as well if they were simply left to their own devices. For all we know, the answer may be the latter.

Back in the late nineties, Ellen Galinsky, the president and co-founder of the Families and Work Institute, had an inspired idea. Rather than blithely speculating about how children experience their parents' efforts to balance work and home, she decided to ask them directly. Her organization did a detailed, comprehensive survey of over 1,023 kids, ages eight to eighteen, and in 1999 she published and analyzed the results in *Ask the Children: What America's Children Really Think About Working Parents.* The data were quite clear: 85 percent of Americans may believe that parents don't spend enough time with their kids, but just 10 percent of the kids in Galinsky's survey wanted more time with their mothers, and just 16 percent wanted more time with their dads. A full 34 percent, however, wished their mothers would be "less stressed."

Maybe dinner should be the new family dinner.

chapter five

adolescence

They don't tell you, when you become a parent, that the hardest part is way, way down the road. —Dani Shapiro, *Family History* (2003)

IT'S A WARM EVENING in Lefferts Gardens, one of those pretty Brooklyn neighborhoods that was still affordable to middle-class New Yorkers before the city's real estate boom, and six mothers, all interconnected through the usual ties (work, kids, community groups), are clustered around a kitchen table in an old brownstone, discussing their adolescents. Their conversation is lively and somewhat mischievous, but not all that surprising at first. Then Beth, a public school teacher and the youngest of the lot, mentions that her fifteen-year-old, Carl, has lately "been using his intelligence for evil."

The women in the group all stop talking and look at her.

"Instead of getting good grades, he figures out how to get around the administrator," she says, referring to the software she's installed to regulate his computer use. "He's been on Facebook and not doing his homework. And then I see, like, three inputs for 'Russian whore.'"

Or so I thought she said when I first transcribed the tape. When I followed up with Beth sometime later, she informed me that I'd misheard: It was "*three-input* Russian whore."

At any rate, Samantha, who also teaches public school, dives in at this moment with the force of a cannonball. "Take the freaking computer, Beth!" she cries. "Take it!"

"He has to use it. They turn things in online."

"On yours, then," Samantha answers. "Take it, Beth! Take it!"

"Put a desktop in the kitchen," suggests Deirdre, the hostess of the evening. She and Beth work in the same building.

"That's what we did," says Beth. "We put it in the living room." She goes on to explain that her son's buzzing around for porn isn't what's bothering her per se (though she's not thrilled with this *particular* porn, which she thinks is disfiguring his ideas about sex). What's really bothering her is that he's spending far too much time online generally, and he's willfully disobeying her by doing so, and his grades are sliding.

Samantha is not yet appeased. "But if he flunks out of school, Beth, what's going to happen?"

"He's not going to flunk out. He only got one D." Then she pauses and considers. "Though when I called his therapist and said, 'I found hours worth of porn on his computer,' the therapist had no idea. So he wasn't talking about that."

"Yeah, but I've had that too," says Gayle, a substitute teacher, quite suddenly. She has, until now, said little. All heads swing her way. "Mae"—her daughter and the best friend of Samantha's oldest, Calliope—"was in therapy and spent a year's worth of my money not talking to the therapist about the real issue, which is that she was cutting herself. Instead, what they talked about was how much she hated violin."

And so the stories slowly seep out. Kate, who first met Deirdre at the local park when their firstborns were toddlers, says her oldest, Nina, did something this summer that created so much tension between her and her husband that she couldn't even bring herself to discuss it. (I would later learn that it involved a minor shoplifting incident.) While at college earlier that year, Nina had also made the mistake of forwarding a paper to her professor that still contained visible traces of proposed edits by her father.

At this point, Samantha finally gives in. She puts her elbows on the table, bows her head, and rests her brow in her hands. "Everyone's in the same club," she says. "Everyone has the same stories." She looks up at the group. "I mean, please. I have *police* stories."

Police stories? All along, as Samantha's friends have been speaking, I've been under the impression that she's been spared these misadventures and is even a tad scandalized by them. But it turns out to be quite the opposite. She's been identifying with everyone's troubles from the start.

whose transition is it?

When prospective mothers and fathers imagine the joys of parenthood, they seldom imagine the adolescent years. Adolescence is the part of parenting that is famously unfun, the stretch of childhood that Shakespeare dismisses as useless save for "getting wenches with child, wronging the ancientry, stealing, fighting," and that Nora Ephron opined can only be survived by acquiring a dog ("so that someone in the house is happy to see you"). Gone are the first smiles, warm nuzzles, and cheerful games of catch. They've been replaced by 5:00 A.M. hockey practices, renewed adventures in trigonometry (secant, cosecant, *what the—*?), and middle-of-the-night requests for rides home. And these are the hardships generated by the *good* adolescents.

But here's the truth of the matter. The children of those women at Deirdre's table? Also the good adolescents. Almost all attend either very good universities or competitive New York City public high schools (one child attends a private high school, also good); all have well-developed interests and talents outside of school. All, in person, come across as self-confident, thoughtful, and considerate.

Yet their parents are still going half-mad. Which raises an important question: Is it possible that adults experience adolescence differently from children? That the category, in fact, might be more useful for *parents* than the children it attempts to describe?

Laurence Steinberg, a psychologist at Temple University and quite possibly the country's foremost authority on adolescence today, thinks there's a strong case to be made for this idea. "It doesn't seem to me like adolescence is a difficult time for the kids," he tells me. "Most of them seem to be going through life in a very pleasant haze. It's when I talk to the *parents* that I notice something. If you look at the narrative, it's 'my teenager who's driving *me* crazy.'"

In the 2014 edition of his best-known textbook, *Adolescence,* Steinberg debunks the myth of the querulous teen with even more vigor. "The hormonal changes of puberty," he writes, "have only a modest direct effect on adolescent behavior; rebellion during adolescence is atypical, not normal; and few adolescents experience a tumultuous identity crisis."

For parents, however, the picture appears to be a good deal more complicated. In 1994, Steinberg published *Crossing Paths,* one of the few book-length accounts of how parents weather the transition of their firstborns into puberty, based on a longitudinal study he conducted of over two hundred families. Forty percent of his sample suffered a decline in mental health once their first child entered adolescence—nearly one-half of the mothers and one-third of the fathers. Respondents reported that they experienced feelings of rejection and low self-worth; that their sex lives declined; and that they suffered increases in physical symptoms of distress, including headaches, insomnia, and lousy stomachs. It may be tempting to dismiss these findings as by-products of midlife rather than the presence of teenagers in the house.

But Steinberg's results don't seem to suggest it. "We were much better able to predict what an adult was going through psychologically," he writes, "by looking at his or her child's development than by knowing the adult's age."

Which is to say that a mother of forty-three and a mother of fifty-three have far more in common, psychologically speaking, if they both have fourteen-year-olds than two moms of the same age with kids who are seven and fourteen. And the mothers of the adolescents, according to Steinberg's research, are much more likely to be experiencing distress.

Steinberg has a theory about why this is. Adolescents, in his view, are the human equivalent of salt, intensifying whatever mix they're in. They exacerbate conflicts already in progress, especially those at work or in the marriage, sometimes unmasking problems parents hadn't recognized or consciously acknowledged for years. Steinberg might even go so far as to say that the so-called crises of midlife would be a good deal less troublesome if adolescents weren't around. But teenagers have an uncanny way of throwing problems, whatever they are, into high relief.

All children do this, of course, to some degree. The question is, why do adolescents have this effect more than, say, children of seven? For that, a historical explanation is useful, and it would run something like this: adolescence, more than any other phase of child-rearing, is when the paradoxes of modern childhood assert themselves most vividly. It is a particularly problematic time for a child to be, as Viviana Zelizer would say, useless.

Adolescence is a modern idea. It was "discovered" by Stanley Hall, the psychologist and educator, in 1904, twenty-eight years *after* Alexander Graham Bell patented the first telephone. This is not to say that adolescence didn't exist before 1904, and that it isn't a physiologically distinct phenomenon, accompanied by discernible biological changes. It most certainly is, and I'll be talking about them in this chapter. But adolescence is a cultural and economic phenomenon too, born of a particular moment in time. It is not an accident that Hall "discovered" adolescence at the exact moment when the nation was becoming sentimental about its young, extending them special protections and keeping them home rather than sending them off to work in the expanding cities and factories. For the first time, parents found themselves protecting and supporting much older children. And their conclusion, after staring at those children at close range during the teen years, was that they had to be going through a terrible period of "storm and stress," as Hall put it. How else could parents explain the chaos they were witnessing?

Yet it could simply be that the advent of the modern childhood—the protected childhood—is especially problematic for older children. Parents today have no choice, of course, but to shelter their children for long stretches of time. Kids are no longer allowed to drop out of school in order to work, and the world now requires more and more schooling to succeed. What's more, parents feel a great *need* to protect their children. Many, especially in the middle class, have waited forever to have them. They fear for their physical safety and economic security. They've been told—by experts, by other parents, by a variety of media—that they ought to spend untold hours nurturing them. Nurturing has become their way of life.

But as children get older, they crave independence, agency, a sense of their own purpose. Keeping them sheltered and regimented for so long, while they're biologically evolving into adults and striving to become who they're meant to become, can have some pretty strange and exhausting consequences. The contemporary home becomes a place of perpetual liminal tension, with everyone trying to work out whether adolescents are adults or kids. Sometimes the husband thinks the answer is one thing while the wife thinks the answer is the other;

sometimes the parents agree but the child does not. But whatever the answer—and it is usually not obvious—the question generates stress.

I SAID IN THE introduction that this is a book about the effects of children on parents. During adolescence, those effects can be especially intense, exposing us in our most vulnerable and existential states. It's not an accident that most parenting blogs are written by mothers and fathers of small children. Part of it, yes, is that these parents are responding to the novelty of their situation. But part of it, too, is that the challenges they're writing about are usually so generic that they're betraying no confidences in revealing them. It does not violate your children's privacy to say they detest peas, and it's not a particularly poor reflection on your parenting either. Whereas writing about adolescents is different. They're incipient adults, with idiosyncratic habits and intricate vulnerabilities; they're unlikely to welcome daily blog posts from their mothers and fathers about their lives. Their parents are no longer inclined to share these stories, either, at least not publicly. The fears that parents harbor are no longer about what foods to feed their children, or what activities they should do. They're about whether their children are moral, and whether they're productive, and whether they're comfortable and sensible and capable of fending for themselves.

Nevertheless, there's clearly a pent-up desire to talk about this stage. When Steinberg first began working on *Crossing Paths* (which he ultimately wrote with the assistance of his wife, Wendy), he identified 270 families who met his requirements. In the end, 75 percent of them agreed to participate. In most social science projects, he notes, the number of willing participants is closer to 30 percent. "We were astounded by the enthusiastic response," he writes in his introduction, "and soon found out the reason why: Parents of this age group are extremely perplexed by the changes they see in their families and are interested in why this period of their lives is so disturbing."

Because this period can be so complicated, I use only first names—and not real ones, but ones supplied by the subjects themselves—in this chapter. It's the only place in the book where I do so. But it seems necessary. In adolescence, children's lives become difficult and messy; there's too much potential harm and not much point in revealing their identities, or that of their parents.

the useless parent

Though she's wearing her workout clothes, you can still make out the hippie that Samantha once was—she's got a gorgeous gray mane of hair, which I hadn't fully noticed at Deirdre's house but can now see in all its glory because she's let it loose from her ponytail following her run. We're sitting in her kitchen in Ditmas Park, one of those miraculous Brooklyn neighborhoods that continually surprises out-of-towners, with its grand stand-alone houses and wrap-around lawns and proper driveways for that rare New York commodity, the family car. Samantha and her husband, both teachers in the New York City public school system (Bruce is also a musician), had the good sense to buy a place here nineteen years ago, when the getting was still cheap by city standards ($234,500) and the neighborhood a bit more diverse. Samantha is African American. Bruce is "the whitest guy ever," according to Calliope, their daughter. Calliope is a fierce beauty, now twenty years old and home for the summer, going into her third year at one of the most competitive colleges in the United States. She joins us at the kitchen table.

"Which bagel?" asks Samantha.

Calliope looks at her with a combination of irritation and affection. "Um, do you know me?" (As in: *How many times have I eaten bagels with you? Hello?*)

Samantha rolls her eyes, grabs one, begins to slice.

The family began calling Calliope "Alpha," as in "Alpha girl," when she was still in high school and was, to put it mildly, very certain about what she wanted. Throughout this brunch, I will hear all sorts of stories about her formidableness. "Calliope would ask for things," says her brother Wesley, lanky and sixteen and mellow as tea, as he strolls in from the living room, "and *get* what she asked for." He's got his guitar with him and begins to strum. He plays guitar and piano and drums with equal dexterity.

"That's not true!" says Calliope, half-laughing, half-appalled.

"Calliope, you just sorta *dominated* the house for a while."

"*I* dominated the house? *Mom* dominated the house."

Perhaps because they both have forceful personalities, mother and daughter clashed a lot while Calliope was still living at home. When I first met her at Deirdre's house, Samantha had recounted one particularly harrowing fight between the two of them, though she never mentioned what started it. So today I ask. Samantha isn't even certain she remembers. Wesley does, and leaps right in. "Well, Calliope had a high school essay due the next day, and a college essay due in a month. So you"—he looks at his mother—"wanted her to work on the college essay, but you"—now he looks at his sister—"wanted to work on the essay due the next day. So you basically said, 'Mom, back off, I need to do this essay tonight.' " He recounts this story with admirable even-handedness. "And you"—Wesley looks at his mother again—"were trying to emphasize your point that the college essay needed to be done."

Samantha waits. But that's it, apparently.

"You just went back and forth like that for a long time," says Wesley. "And then Dad stepped in."

Samantha looks puzzled. "That's so stupid. Why would I not want her to do her essay for the next day?"

Wesley again responds with tact. "Well," he says, "in hindsight, you can understand her perspective. But at the time, you wanted to be heard. Which is why the argument continued."

This argument, like so many arguments, wasn't about much. It was what roiled beneath the surface that clearly upset Samantha. She had ideas about her daughter's priorities, but her daughter had different ideas, and Samantha could feel her authority slipping away. She could even detect a hint of mockery in Calliope's responses to her suggestions. Samantha hates it when she's being mocked.

"The cursing doesn't bother me," she says a bit later on in the discussion, trying to describe what she experiences when her kids swear at her. "It's the *tone*."

"Or when we say 'relax,' " says Calliope. "Or 'chill.' "

Samantha springs up from her chair as if released from a slingshot. "Yes! *Oh my God.*" She starts pacing. "It's so minimizing. Like, 'You're not important.' "

"Well, you *are* really wound up sometimes," says Wesley, mildly. "Like when you remind us for the tenth time the cleaning lady's coming—"

Samantha cuts him off. "That's because I say, 'Remember, she's coming tomorrow,' and you say"—she switches to the lower register of an aggrieved fifteen-year-old boy—"*Relax, Mom, I know what day it is. What the fuck!*" Everyone, including her, is laughing now. "That's what *I* hear."

THE CONVENTIONAL WISDOM ABOUT adolescence is that it's a repeat of the toddler years, dominated by a cranky, hungry, rapidly growing child who's precocious and selfish by turns. But in many ways the struggles that mothers and fathers face when their children hit puberty are the very opposite. Back when their children were small, parents craved time and space for themselves; now they find themselves wishing their children liked their company more and would at least treat them with respect, if adoration is too much to ask. It seems like only yesterday that the kids wouldn't leave them *alone*. Now it's almost impossible to get their attention.

I ran across a remarkably meticulous study from 1996 that managed to quantify the decline in time adolescents spend with their families. It followed 220 working- and middle-class children from the Chicago suburbs, once when they were in grades 5 through 8, and again when they were in grades 9 through 12. At each interval, the researchers spent a week paging these kids at random, asking them to identify what they were doing, with whom, and whether they were having fun. What they found, 16,477 beeps later, was that between fifth and twelfth grades, the proportion of waking hours that children spent with their families dropped from 35 to 14 percent.

Another Brooklyn mother, whose circle of friends overlaps with some of the women at Deirdre's table, likened her fifteen-year-old daughter to race car driver. "I change all of her tires, polish up the car, and get out of the way," she told me. "Then she peels out. I'm the pit crew."

It takes a lot of ego strength to be in the pit crew. It means ceding some power to your children, for one thing—decisions that were once under your purview move to theirs—and it means receding somewhat, accepting that they've recast their lives without you, or your goals, at the center. Joanne Davila, a psychologist at SUNY Stonybrook, puts it this way: "During childhood, it's about trying to help develop who your kid's going to be. During adolescence, it's about responding to who your kid *wants* to be." And that's the generous interpretation, told from the parent's point of view. From the adolescent's, it's often a good deal less rosy. "The adolescent," writes Adam Phillips in his book of essays, *On Balance*, "is somebody who is trying to get himself kidnapped from a cult." Parents go from their kids' protectors to their jailers and are then told repeatedly what a drag this is.

Indeed, one of the most striking—and concretely measurable—ways of seeing how critical kids are of their parents at this stage can be found in Ellen Galinsky's *Ask the Children,* which, as I said in the previous chapter, is based on a survey of over one thousand kids in grades 3 through 12 and ranges over a wide variety of topics. At one point, Galinsky asked her interviewees to grade their parents. In almost every category, seventh- to twelfth-graders rated their parents considerably less favorably than did younger children. Fewer than half of the mothers and fathers were given an A by their older kids on "being involved in their children's education, in being someone whom their children can turn to if they are upset, in spending time talking with their children, in establishing family routines and traditions, in knowing what's going on in their children's lives, and in controlling their tempers." (In fairness, younger children assigned their parents equally bad grades for controlling their tempers.)

Ingratitude is already one of the biggest heartaches of child-rearing. (Shakespeare, famously, from *King Lear:* "How sharper than a serpent's tooth it is to have a thankless child.") During adolescence, that ingratitude is additionally seasoned with contempt. It's a lot to handle, especially for a generation of parents who have made their children the center of their lives. Months after I met her at Deirdre's, Gayle, Mae's mother, mentioned that she could count on two hands the number of times she had left her daughters with a babysitter when they were young. Gayle and her sisters, on the other hand, were left with babysitters for two weeks at a clip when they were growing up. "And you know what?" she said. "We were *happy.*" But she wanted to be more involved as a mother. She wanted to be *present.* And so she was. Then adolescence hit, and her girls got old enough to use the New York City subway unaccompanied; Mae, her oldest, got prickly around her, and their conversations became increasingly fraught. Gayle's intensive involvement with her children didn't inoculate her from rejection, or from any of the pain.

CONTENDING WITH THE TENSION of separation during adolescence is hard enough. What makes it even harder, in Steinberg's estimation, is the contrast to the bonded, reasonably tranquil period preceding those years. A number of psychologists have pointed out that adolescence creates a dramatic discontinuity in the entire family system, destabilizing dynamics, rituals, and a well-maintained hierarchy that has been in place throughout most of elementary school. The *Blackwell Handbook of Adolescence* goes so far as to say that adolescence "is second only to infancy" in terms of the upheaval it generates. Power must be renegotiated. The family must realign. Rites must be reviewed. The "old script," Steinberg writes in *Crossing Paths,* "no longer fits the new characters."

In this way, the challenges of adolescence *are* a replay of the early years: first there was order; now there isn't. And it's not just the dramas of infancy that repeat themselves, but those of the toddler years too: once again, the child is struggling for autonomy, but this time with more reasoning skills and the physical means to carry out his or her plans. Steinberg floats a hypothesis in *Crossing Paths* that is both subtle and daring: "I believe that we have underestimated the positive feelings parents derive merely from being able to physically control their children when they are younger. I do not mean this in a negative sense. The physical power they hold over their children reaffirms parents' sense of control and importance."

Yet parents of adolescents have to learn, by stages, to give up the physical control and comfort that was once theirs. In the end, they are left only with words. This transition is almost a certain recipe for conflict. There's so much yelling, suddenly, and so much (seemingly) gratuitous defiance; simple requests to do work or pick up clothes "lead into temper tantrums," as another Brooklyn mother, a government lawyer, put it. "Ask him, and my son just flies off the handle." While not all researchers agree that adolescents fight *more* than younger children, almost all concur that they fight with more vehemence and skill, arguing most intensely with their parents between eighth and tenth grades. (This is, in fact, precisely the conclusion of a 1998 meta-study that took into account thirty-seven different surveys of conflict between parents and adolescents.) Children at this stage are better able to reason too, and to turn their parents' own logic against them in potentially ugly ways; as any parent of a teenager will tell you, an adolescent knows just what hurts.

"I remember finding out very quickly in high school the few things I could say that would really get to my mom," Calliope confesses at one point.

Samantha shoots her an incredulous look: "Did you? That's so *mean*—"

In her work, Nancy Darling offers a nuanced analysis of what, precisely, makes the adolescent struggle for autonomy so contentious. Most kids, she notes, have no objections when their parents try to enforce moral standards or societal conventions. *Don't hit, be kind, clean up, ask to be excused*—all this is considered fair game. The same goes for issues of safety: kids don't consider it a boundary violation if they're told to wear seat belts. What children object to are attempts to regulate more personal preferences, matters of taste: the music they listen to, the entertainments they pursue, the company they keep. When children are young, these personal preferences don't tend to cause parents too much anxiety because they're benign in most cases. Barney? Annoying, but unobjectionable. That little boy across the way? A little rowdy, but a decent kid. The Jonas Brothers? Cloying, but a little syrup never hurt anyone.

The problem, says Darling, is that during adolescence questions of preference start to bleed into questions of morality and safety, and it often becomes impossible to discern where the line is: *That kid you're hanging out with? I don't like how he drives or the stuff he's introducing you to. Those games you're playing? I don't like all the violence and disgusting messages they're sending about women.* Even an issue as banal as wearing jeans to church, Darling writes in one of her blog posts on this subject, is a humdinger. Is that a matter of personal expression? Or an outrageous violation of social custom?

And it's often the banal issues of taste that become the most explosive. At her kitchen table, Samantha tells me about a tiff she's recently had with Calliope over the merits of Beyoncé. Or perhaps, more accurately, one should call it a *misunderstanding* over the merits of Beyoncé. Samantha mistakenly thought the pop singer and actress was someone else who was much more down-market. She mentioned to Calliope that she couldn't fathom why Calliope adored such an unrefined human being.

This snooty (and, it would turn out, misinformed) point of view seriously annoyed Calliope, making her wonder what particular type of bee had crawled into her mother's bonnet. Who she listened to was her business, her prerogative. But to Samantha, this was a quasi-moral issue: she thought that Beyoncé represented the wrong kinds of values, and she was dismayed, in her heart of hearts, that her kid admired this person.

Then, as they were talking, Samantha discovered that she wasn't even thinking of the right woman. Her daughter showed her a picture of Beyoncé online, and Samantha immediately realized she was confusing her with someone else. This just made the argument doubly maddening to her daughter, and doubly mortifying to Samantha—and doubly moot.

IN HIS RESEARCH, STEINBERG finds that parents' experience of their children's adolescence can be exacerbated by any number of factors. One is being divorced: there is a big mental health differential between married parents and divorced ones as their kids enter puberty. Steinberg suspects one of the reasons is that the relationship between a divorced parent—a mother in particular—and her child can be so intense that it's hurtful when the child starts to separate.

Steinberg has also found that parents of a child of the same sex weather the pubescent years far worse than parents of a child of the opposite sex. (The conflicts between mothers and daughters, he added, are especially intense—a finding that has been duplicated over and over again by researchers other than Steinberg.) He speculates that their difficulties may again be explained by an abrupt break in equilibrium: before adolescence, parents tend to be much *closer* to their child of the same sex, which makes that child's efforts to separate all the more painful.

There is, however, another possible explanation for this phenomenon, one I ran across not infrequently in interviews: having a child of the same sex opens up an uncomfortable opportunity for identification. The child, now older, reminds the parent of himself or herself, or who he or she was in high school. "I think it's a lot easier to parent a child before their struggles start to reflect your struggles," says Brené Brown, a researcher at the University of Houston who specializes in thinking about shame. "The first time our kids don't get a seat at the cool table, or they don't get asked out, or they get stood up—that is such a shame trigger."

Even more complicated, the child can represent threatening

teenagers from their parents' own youth. Samantha brought this up at Deirdre's house when she was talking about how intimidating she sometimes found Calliope. "Sometimes I look at my child," she told the group, "and I'm so frightened of her because of who that person was in high school. You go back to who you *were* in high school. You have to remind yourself: *Wait a second. I'm the parent here.*"

In addition, Steinberg has found that adolescence is especially rough on parents who don't have an outside interest, whether it be work or a hobby, to absorb their interests as their child is pulling away. In his sample, this was true, strangely, whether the parent was an involved parent or a disengaged one, a helicopter or a remote-controlled drone. "The critical protective variable was not, as some might expect, whether or not an individual invested a great deal in parenting," he writes. "It was the *absence* of non-parental investment." Mothers who'd made the choice to stay home were especially vulnerable to a decline in mental health. But so were parents without hobbies, and so were parents who didn't find fulfillment in their jobs and viewed them more as a source of pay than a source of pride. It was as if the child, by leaving center stage, redirected the spotlight onto the parent's own life, exposing what was fulfilling about it and what was not.

This is nowhere more evident to me than when I sit down to talk to Beth, the public school teacher who complained at Deirdre's house that her fifteen-year-old was "using his intelligence for evil." She already has one daughter in college. And that daughter? Amazing. Her adolescence passed without much drama; Beth is plainly awed by her. But her son Carl . . . his passage to adolescence was another story entirely. The porn she could handle (what teenage boy isn't curious about sex?), but the defiance, the cursing her out, the endless hours he sank into *StarCraft*—all of that wore her down in a terrible way. She found it worse than wearying, actually: it was confidence-killing, as if she were doing things all wrong. He was a bright kid, testing into a competitive public middle school and high school, but she could see he was struggling academically, and struggling more generally with issues of motivation and initiative. Everything became a tiresome contest of wills; everything devolved into a fight. "It seemed like whenever we left to go anywhere," she says, "to get him out of bed, there'd always be a huge argument." Her moods became tethered to his. On a week when their relations were warm, she'd feel better; if he pulled away, even though there'd be less arguing, "I'd get depressed."

And then came the summer between his freshman and sophomore years, when he got so slothful he began to sleep on a bare mattress. "I'd say to him, 'Carl, just get up and put a sheet on your bed.' And he'd say, 'Get out of here. You've failed as a mother.' "

In late August, she gave him an ultimatum: respect the rules of the house, or go live with Dad. So he left. Before that, he'd spent almost all his evenings, all his weekends, all his *life* practically, with her.

"And it was kind of like, *What's my purpose?*" she says. "I never thought about my job or career as something I was *devoted* to. My kids had always been my number one."

Now, with Carl at her ex-husband's house and her daughter headed back off to college, she realized she wasn't looking forward to the new school year at all. "If all I had was my job," she says, "I wanted something else."

But she had a constructive response to this distress. She wrote a letter to Carl, and one to her ex-husband, too, trying to relate, trying to empathize, trying to accept the blame that was hers. She took Carl to a psychiatrist and got him a proper evaluation, which yielded, in the end, a common enough diagnosis—ADHD—for which there was a common enough treatment—medication. His grades went up by one or two letters each. After visits with her, he started leaving her voice-mail messages like this one:

Hey, Mom, it's Carl. I just wanted to say I had a really great time today and thank you a lot. I know that I was like, uh, you know, I was, um, very irresponsible with the laptop . . . overall, throughout life. I was, you know, I was a very hard child. I just want to say I'm very sorry, and thank you for putting your faith in me once again, even though we've had so much trouble in the past. I really love you. . . .

He left that one six months before we sat down to talk. "I'm saving it *forever*," she says, after playing it for me.

She didn't pressure Carl to come home and live with her. And he didn't. Their relationship remained tenuous, easily capsized. But Beth began to detect a subtle shift in her attitude toward her job. By winter, she realized: she really *liked* her students. There were a couple in particular who really moved her and who got really attached to her—a boy who wanted her to come see him act in a play and a remarkably resilient girl whose mother had died. "I was getting from my students what I wasn't getting from my son," she says. "Appreciation, connection. But I had to get to the point where I could recognize that—realizing that I can't get everything from my family, my kids."

marital strain, part ii

"There was a recent issue where we strongly, *strongly* disagreed," Kate is saying, "and I was right."

Her husband Lee, a man in his midfifties with longish gray hair, gives her a baffled look. "I don't even know what issue you're referring to."

"The party at Paul's."

Lee sucks in his breath. "But that's where—"

"Let me talk, okay? I feel strongly about this." Lee stifles his frustration. He yields the floor.

It's a tense moment. Kate and Lee have been together for twenty-two years, and their marriage is solid: they exercise together, shop together, and have all of their dinners together; they both work at home and have managed to keep the peace. But when their son and daughter entered adolescence—the kids are now fifteen and nineteen, respectively—Kate noticed a transformation in their marital dynamics. She said it outright, the night we were all sitting around Deirdre's kitchen table: "There's a lot more discord between us, having teenagers around. I'm just assuming that when they're both out of the house, there'll be a lot less."

This morning, at their home just around the corner from Deirdre's place, Kate and Lee are talking about that discord, or at least generously trying. It's hard.

"If the kids go to a party at somebody's house," Kate resumes, "I want to know that there's going to be a parent present. And if they can't tell me, I will call and find out." She means it. She's done it. "And this time," says Kate, "I let it slip a little bit, because we were dealing with a friend of Henry's who we've felt was very trustworthy before."

So her son went to the party, and the parents were away. "It was one of these things where the kid lied," says Kate. "He told his parents he'd be sleeping at somebody's house, but instead he invited everybody over in his grade, and the police showed up." When the parents returned, they were mortified, sending emails of apology to all the families involved and making their son personally call everyone and say the same.

So what, I ask, was Kate and Lee's argument about?

"Whether he should have been allowed to go," says Kate. "Lee didn't think it was as big a deal."

"Which *remains* my view," says Lee.

"It shouldn't," says Kate. "If *we* left the house, and there was a party, and the police came, and *our* house was trashed, that would have been a nightmare. I don't want my kid to be party to that."

IF ADOLESCENTS ARE MORE combative, less amenable to direction, and underwhelmed by adult company, it stands to reason that the tension from these new developments would spill over into their parents' marriages. But gauging the influence of teens on relationships is a tricky business. A lot of confounding factors—career dilemmas, health troubles, the routine difficulties of coping with aging parents—can sneak into the mix, making it hard to distinguish between the effects of adolescent children and other challenges common to midlife; it's also not uncommon for marital satisfaction to steadily decline over time from sheer habituation. (Certainly, sexual frequency in married couples declines over time.) But that hasn't stopped some researchers from trying to measure the impact of adolescents anyway, and a number have concluded that marital satisfaction levels do indeed drop once a couple's firstborn child enters puberty—on *top* of a more general decline in marital satisfaction.

In fact, many studies will go to elaborate lengths to show just how the onset of puberty and a decrease in marital happiness coincide. A 2007 survey published in the *Journal of Marriage and Family* went so far as to track the "growth spurt[s], growth of body hair, and skin changes" of the children of its 188 participating families—as well the voice changes in boys and the first menses in girls—in order to see if marital love and satisfaction levels dropped even more precipitously as these changes occurred. They did.

This strife is by no means preordained. There are couples who will tell you that they've reclaimed their evenings and resumed adult conversation since their children hit puberty, that they interact almost as they did before their kids came along. Thomas Bradbury, a marriage researcher at UCLA, likes to point out that if a couple has withstood their first child's passage to adolescence, the parents are "survivors," with a far more durable marriage than average: "They have weathered a lot of storms, and are settled into routines that work for them, more or less." But overall, the evidence, both in research labs and in clinical settings, seems to be that relationship dynamics are stressed, rather than strengthened, by adolescence. Andrew Christensen, a UCLA professor who both does research on couples therapy and has a clinical practice—therefore experiencing family conflict daily, *in vivo,* and not just on paper—gives a perfect example of the kind of more subtle conflict he sees among parents of adolescents:

> Inevitably, we see ourselves in our kids. And then we see our partner acting toward our child the way our partner acts toward *us.* Like, let's say Mom is upset with Dad because he hasn't been very ambitious—he's a little lazy, hasn't made it in the world the way he should have. And then she sees her adolescent son showing similar qualities of not taking initiative. She might be angry at the dad for not being a better role model for him, and fears he may be turning into a slacker too. But if you take Dad's point of view, he sees Mom being critical of the son the way she's critical of him, and he's protective of the son. *That tends to be one of the worst-case scenarios of parenting conflict we see clinically* [emphasis mine].

Long gone are the days when the fights started with, "I got the baby *last* night," or, "What were *you* doing all day?" Either directly or indirectly, the fights increasingly revolve around who the child is, or is becoming. Projection is now possible. Identification is now possible. Which means that competitiveness, envy, disgust—all are possible, all can rear their heads. These aren't feelings evoked by younger children. They're brought on by other *adults.*

Mistaking teenagers for adults can be especially problematic in high-conflict relationships. As children mature and develop the capacity to reason and empathize, it's increasingly tempting for their parents to recruit them in their arguments, which only aggravates matters. *Now you're dragging Charlie into this?* (In one intriguing study, teenage girls felt more pressure to side with their mothers if their parents were still married, while teenage boys felt more pressure to do so if their parents were divorced—suggesting, perhaps, that teenage sons feel compelled to step in as their mothers' protectors if their fathers are no longer at home.)

In *Crossing Paths*, Steinberg gives another example of how identifying with one's adolescent can strain a marriage. I should say up front that I haven't seen this finding replicated. (Then again, I'm not sure anyone's bothered to try.) Steinberg noticed a substantial decline in the marital satisfaction of his male subjects when their teenagers began to date. "In fact," he writes, "the more frequently the teenager dated, the more unhappily married the adolescent's father became." If his teenager was a son, Steinberg noticed, the effect was especially bad. He surmises it has to do with a combination of sexual jealousy and nostalgia for a lost era of open-ended possibilities. But he admits he didn't quite think it was possible to put this question to his subjects directly.

WE ALSO SHOULDN'T UNDERESTIMATE the effect that adolescents can have on relationships simply because they introduce new subjects over which to disagree. Before their children are born, parents don't tend to discuss what their policies will be toward, say, dating, or wearing short skirts, or staying out late. "At least there are coaches for breast-feeding and sleeping," says Susan McHale, a developmental psychologist at Penn State. "But then along comes adolescence, and you don't know what to expect or how to handle it."

Especially if the child screws up. "One parent is the softie, and the other's the disciplinarian," says Christensen. "That comes up a lot, and it's a very big challenge. Dad sharing his recollections with drugs and alcohol, but Mom remembering something bad happening. And then they divide over it."

This is the kind of argument that Kate and Lee seem to have a good deal. At their son's soccer game, they finally tell me about the fight they had when their daughter, Nina, attempted to shoplift a skirt, a little experiment for which the family paid a steep fine. Both agree that the circumstances driving their daughter to such an act were unusual, and that her behavior was atypical—it was summer, she had just graduated from high school, and she was living in a strange city, where she knew virtually no one. But their reactions at the time were very different. Kate was so angry that she refused to pick up the phone when her daughter called to discuss it. Whereas Lee went out of his way to console her.

"As usual," says Lee, "I didn't lay into her with fire and brimstone. I could hear how terrible she felt. Rather than reacting to the deed, I was hearing *her*."

"But my problem with it," explains Kate, "was the same as the term paper." She's referring to the paper Lee edited that still contained traces of his editing suggestions when her daughter handed it in to her professor. "When you do it on a certain level and you get caught," she says, "it becomes a crime that can affect your *life*. They have to really, really know that it's really, really serious."

"But it wasn't plagiarism."

"Yes, but he was *accusing* her of plagiarism, and just accusing her could have cost her her scholarship."

"Well, okay, fine, but, whatever—"

"*No*," says Kate. "Not 'okay, fine.' That's $20,000 a year. So. No."

Or consider this exchange, from Deirdre's table:

KATE: I'm really, *really* strict with the kids, and he knows that I am, so he's totally not. We just had a fight about it today.

They'll go to him to tell him stuff they're afraid to tell me. And he'll say it's okay and tell a funny anecdote.

SAMANTHA: That's my complaint about my husband too. He minimizes stuff.

BETH: Same here. I make the rules, and he's the friend.

This idea seems to be suggested by data too—particularly in a large, renowned longitudinal study by the University of Michigan, parts of which have been ongoing since 1968. In a fairly recent sample of nearly 3,200 parents of ten- to eighteen-year-olds, a disproportionate share of mothers said that the task of discipline fell to them alone (31 percent, versus just 9 percent of the dads). Mothers also reported setting more limits for their adolescents: they were 10 percent more likely than dads to set limits on video and computer games; 11 percent more likely to set limits on what types of activities they did online; and 5 percent more likely to regulate how many hours of television their kids could watch per day.

For the last decade or so, says Nancy Darling, research—including her own—has shown pretty consistently that adolescent girls and boys *both* direct more verbal abuse at their mothers than at their fathers, and they make more physical threats against their mothers too (though both boys and girls are more likely to enact their aggressive impulses against their fathers). According to Steinberg's research, mothers are also more likely than fathers to quarrel with their adolescent children, and (perhaps as a result of this high-frequency conflict) to bring more family stress into their workplaces.

These complicated dynamics may explain why mothers, contrary to conventional wisdom, tend to suffer less than fathers once their children have left the home. Kate readily admits her relationship with Nina improved considerably once she went off to college. As Steinberg concisely puts it: "Women's personal crises at midlife do not come from launching their adolescents but from living with them." Mothers are more attuned to their children's separation process all along—quarreling with them more, being on the receiving end of more scorn and slights. Whereas to fathers, their children's departures from home feel more abrupt and occasion more questions and regrets.

the adolescent brain

As Wesley was assessing the conflicts between his sister and his mother, I thought I could clearly discern his self-appointed role within the family. He was the peacemaker and the diplomat, the kid who made a scrupulous point of not being difficult and not making waves. That's what you do with a strong mother and older sister: keep your head down and your hat brim pulled low.

Yet it was Wesley, sensitive Wesley, so self-controlled and empathetic in every way I could see, and so talented in ways that would make any parent flush with pride, who got dragged home by the police at 4:00 A.M. Calliope was about to graduate; the family was expecting Samantha's mother-in-law the following day; they already had a guest in the spare room. Into this picture entered Wesley, tucked into the back of a squad car. He and his friend had been out "egging"—tossing eggs at windows of homes in the neighborhood.

"We hadn't considered the fact that it was a misdemeanor," says Wesley as we're finishing up our bagels. He's calm as he describes the incident. "We just did it because it was fun."

That is not, needless to say, how his mother views it. "One of those houses had a kid who Wesley played baseball with," says Samantha. "He didn't know."

Wesley exchanges a discreet glance with his sister. He knew.

There were other things Samantha hadn't known either, apparently, until we had this conversation. Like how Wesley got out without her knowledge. He'd wait until his parents were sleeping and the fans were running loudly. Then he'd tiptoe downstairs. After a while, his methods became even more advanced. "I started to hop off the roof," he explains to her, with serene matter-of-factness. "And then it was impossible for you to track me."

"Wait." Samantha does a classic double take. "What roof?"

"*The roof.* I would climb out my window and hop off the roof. And then climb back up when I got home."

Samantha stares at him, saying nothing. Then: "How did you get back in?"

"I'd climb," he says. "There's a big thing on the railing I could step on. I didn't think I could do it until I tried. That made everything a lot easier. Before, it was a lot of work for me."

TEENAGERS MAY STRIKE US as precocious grown-ups one minute, but only one minute later we realize that they are not. Their forays into independence can tip easily into baffling excess, as if they're experimenting not just with notions of autonomy and self-determination but with their own mortality—and along the way, the mercy and forgiveness of the law.

As with the mysterious behaviors of toddlers, this conduct has distinct neuronal underpinnings. As recently as twenty years ago, researchers hadn't given much thought to the teen brain, assuming that adolescents were essentially adults with rotten judgment. But recently, with the advent of magnetic resonance imaging to more closely examine brain topography and function, researchers have discovered that adolescents do not walk around with a defect that prevents them from properly assessing risk. B. J. Casey, a neuroscientist at Weill Medical College of Cornell University, notes that it's just the opposite: adolescents *overestimate* risk, at least when it comes to situations involving their own mortality. The real problem is that they assign a greater value to the *reward* they will get from taking that risk than adults do. It turns out that dopamine, the hormone that signals pleasure, is never so explosively active in human beings as it is during puberty. Never over the course of our lives will we feel anything quite so intensely, or quite so exultantly, again.

An overstock of dopamine is inconvenient enough. Then consider that the prefrontal cortex, the part of the brain that governs so much of our higher executive function—the ability to plan and to reason, the ability to control impulses and to self-reflect—is still undergoing crucial structural changes during adolescence and continues to do so until human beings are in their mid- or even late twenties.

This is not to say that teenagers lack the tools to reason. Just before puberty, the prefrontal cortex undergoes a huge flurry of activity, enabling kids to better grasp abstractions and understand other points of view. (In Darling's estimation, these new capabilities are why adolescents seem so fond of arguing—they can actually *do* it, and not half-badly, for the first time.) But their prefrontal cortexes are still adding myelin, the fatty white substance that speeds up neural transmissions and improves neural connections, which means that adolescents still can't grasp long-term consequences or think through complicated choices like adults can. Their prefrontal cortexes are also still forming and consolidating connections with the more primitive, emotional parts of the brain—known collectively as the limbic system—which means that adolescents don't yet have the level of self-control that adults do. And they lack wisdom and experience, which means they often spend a lot of time passionately arguing on behalf of ideas that more seasoned adults find inane. "They're kind of flying by the seat of their pants," says Casey. "If they've had only one experience that's pretty intense, but they haven't had any other experiences in this domain, it's going to drive their behavior."

Over time, researchers who look at the adolescent brain have therefore alighted on a variety of metaphors and analogies to describe their excesses. Casey prefers Star Trek: "Teenagers are more Kirk than Spock." Steinberg likens teenagers to cars with powerful accelerators and weak brakes. "And then parents are going to get into tussles with their teenagers," says Steinberg, "because they're going to try to *be* the brakes."

It's a dicey business, being someone's prefrontal cortex by proxy. But resisting the impulse to be a child's prefrontal cortex takes a great deal of restraint. It means allowing that child to make his or her own mistakes. Only through experience can a teenager—or anyone, really—learn the painful art of self-control.

Complicating matters, adolescent brains are more susceptible to substance abuse and dependence than adult brains, because they're making so many new synaptic connections and sloshing around with so much dopamine. Pretty much all quasi-vices to which human beings turn for relief and escape—drinking, drugs, video games, porn—have longer-lasting and more intense effects in teenagers. It makes acting out especially tempting to them, and it makes their habits especially hard to break. "I used to think that if I locked up my son until he was twenty-one, I'd be okay," says Casey. "But the brain does not mature in isolation. Teenagers are learning from their experiences—the good, the bad, and the ugly."

If it's of any comfort, B. J. Casey and her colleagues speculate that there's an evolutionary reason why Kirk rather than Spock so often emerges the victor in the quest for control over an adolescent's mind. Human beings need incentives to leave the family nest. Leaving home is dangerous; leaving home is hard. It requires courage and learning lessons of independence. It may even require a purposeful recklessness.

In a piece he wrote for *National Geographic* about the teen brain in 2011, the science writer David Dobbs began with his own personal experience, recalling the time his oldest son, then seventeen, had been pulled over by the cops for driving 113 miles per hour. One of the strangest parts of the whole episode, he writes, was when he realized that his son didn't accidentally drive like a maniac. He planned it out. He *wanted* to drive 113 miles per hour. He in fact found it maddening that the cops gave him a citation for reckless driving. " 'Reckless' sounds like you're not paying attention," he told his father. "But I was."

And when I asked Wesley why he was tossing eggs at his neighbors' houses, he too responded just as matter-of-factly, and with a similar explanation.

"Because I wanted to do it," he answered. "There's no rhyme or reason."

And why this particular activity?

He looked at me with some amusement. Only an adult requires logic in this situation. It's the Vulcan's prerogative, logic. Wesley didn't have nearly as much use for it.

"Spontaneous," he answered. "Impromptu. Throw an egg. That looks like fun."

the useless adolescent

Here's a historical point to consider: it's possible—just possible—that adolescents would be less inclined to throw eggs at houses, drive 113 miles per hour on highways, and indulge in all other manners of silliness if they had more positive and interesting ways to express their risk-taking selves. That's the theory of Alison Gopnik, the Berkeley psychologist and philosopher: that modern adolescence generates an awful lot of "weirdness" (her word) because our culture gives older children too few chances to take constructive, and tangibly relevant, risks.

She's not the first to make this point. During the 1960s, Margaret Mead complained that the sheltered lives of modern adolescents were robbing them of an improvisational "as-if" period during which they could safely experiment with who they'd ultimately become; the result of this deprivation was a lot of acting out. Recently, in an interview on NPR, Jay Giedd, who researches the teen brain at the National Institute of Mental Health, put it very well: "These Stone Age tendencies are now interacting with modern marvels, [which] can sometimes not just be amusing anecdotes, but can really lead to more lasting effects." Like driving motorcycles without helmets, standing on top of moving subway trains—all the horror stunts kids do because they think, rightly but tragically, that they'd be fun.

Without romanticizing or overstating the advantages of the past, it is worth noting that there once were more purposeful outlets for adolescents' restless energies. At the beginning of our republic, writes Steven Mintz, "behavior that we would consider precocious was commonplace." He mentions Eli Whitney, who opened his own nail factory before going to Yale at sixteen, and Herman Melville, who dropped out of school at twelve to work "in his uncle's bank, as a clerk in a hat store, as a teacher, a farm laborer, and a cabin boy on a whaling ship—all before the age of twenty." George Washington became an official surveyor for Culpepper County at seventeen and a commissioned major in the militia at twenty; Thomas Jefferson lost both parents by the age of fourteen and entered college at sixteen. "The mid-eighteenth century," Mintz writes, "provided many opportunities for teenagers of ambition and talent to leave a mark on the world."

But by the twentieth century, with improvements in mortality rates, more parents survived to shelter their children, and more children survived infancy. Families got smaller. The Progressive era ushered in a much more humane brand of politics, and laws were passed that forbade many forms of child labor and made public school mandatory and universal. (Between 1880 and 1900, the number of public schools in the United States increased by 750 percent.) These were all positive developments; no humane person pines for the Dickensian child-labor practices of years past. But even at the time, there were liberal social critics who wondered whether these new laws would inadvertently rob children of their courage and independence. In the December 1924 issue of *Woman Citizen*, a writer dared ask whether "Lincoln's character could ever have been developed under a system that forced him to do nothing more of drudgery than is necessitated by playing on a ball team after school hours." And *Woman Citizen* was not a publication that harbored hostility to the Progressive cause. It published an essay version of Margaret Sanger's "The Case for Birth Control" that same year.

No matter. After World War II, older children no longer played a central role in the workforce. The divergent paths to American adulthood converged into a single superhighway, with almost all kids zooming at the same speed through the same program: public school, kindergarten through grade 12.

One could argue that school provides an opportunity for adolescents to take risks, but it'd be a stretch. (If anything, it's easier to make that case about after-school activities—sports, say, or musical theater—than about school itself.) Not all kids are good at school; students excel at different rates and in different subjects; American schools, with their teach-to-the-test mania and standardized curricula, don't accommodate differences particularly well. Today schooling is so rigidly structured, and so painfully regimented, that there's almost no room for flexibility—much less risk-taking—at all.

One could more credibly argue the opposite point: that the vanishing of the "as-if" period explains the sudden emergence of what sociologists now call "emerging adulthood"—that late-blooming time when young college graduates live in shared quarters and surf from one low-paying job to another, figuring out where they want to live and what they want to do. This phase has become the new as-if period, the new time of safe experimentation. Some critics call this period "extended adolescence." But that's really not what it is at all, if you think about it: this so-called "emerging adulthood" is really adolescence in earnest, the first time children have a chance to experiment and find themselves, which they once did far earlier as a simple matter of custom, a matter of course.

SOMETHING ELSE BEGAN TO happen once children were synchronized and sequestered. They started to develop a culture of their own. This culture was only made more powerful by mass media and the advertising age, which also started to boom after World War II. A commercial market exploded around teens, who began to drive trends in popular culture. It's not a coincidence that the word "teenager" emerged in the American lexicon during the forties, and made its first print appearance in 1941, in both *Popular Science Monthly* and *Life.* ("They live in a jolly world of gangs, games, movies, and music," proclaimed the latter.) This was the same moment that both modern childhood and mass media were born. "Teens, for the first time, shared a common experience and could create an autonomous culture," Mintz writes, "free from adult oversight." High schools themselves became the focus of sociological scrutiny. *The Adolescent Society,* a portrait of high school culture in the Midwest, became a classic of sociology in 1961.

As the twentieth century progressed, in short, a paradox evolved. The more time children spent in each other's company, the more powerful their independent culture became; the more powerful that culture became, the less amenable adolescents were to the influence of their parents. Yet even as adolescents were leading increasingly separate lives, and even as they were chafing under their parents' influence, they found themselves increasingly dependent on their parents for resources (cars, money), emotional support, and connections in an increasingly complicated world.

The result was the modern teenager, a class of human beings simultaneously excoriated for being too obstreperous and too helpless. They were likened, all at once, to wild horses and penned veal. The Austrian-born psychologist Bruno Bettelheim probably said it best in the 1970s: "We know so much better what makes them tick, and are so much less able to live with them."

There are ways in which this gap has narrowed in recent years. As ferociously protected as American adolescents may be, their world is more diverse than their parents' was, and more filled with nontraditional family arrangements. The Internet has introduced them to more sex, violence, and real-life horrors (celebrity sex tapes, dismembered corpses from terror attacks, the hanging of Saddam Hussein) than any preceding generation was exposed to, and today's adolescents are more aware of the world's financial instability, with fewer of them coming from the traditional middle class. For their part, parents, as aggressively protective as they've been, are more inclined toward openness than their parents were, being veterans of sex, drugs, and rock-and-roll themselves. Though they may not be as fluent in consumer and pop culture as their kids, they're surrounded by the same electron cloud: they've all read *The Hunger Games* and watched *Friday Night Lights.* "In other words," writes Howard Chudacoff in his conclusion to *Children at Play,* "children's 'aspirational age' has risen, while that of adults has fallen. An eleven-year-old no longer asks for a stuffed animal or fire truck and instead desires a Madden NFL football game, a cell phone, an iPod, or a Beyoncé Knowles CD, while a thirty-five-year-old may also indulge by buying a Madden football game, a cell phone, an iPod, or a Beyoncé Knowles CD."

But there are also important ways in which the paradox of the adolescent's existence—as both helpless dependent and impertinent rebel—has intensified.

Why? Because today's adolescents, more than ever before, are full-time professional students in highly structured environments that keep them at home and dependent on the family purse, seemingly forever. And their parents, having waited so long to have them and having made them the center of their lives, spend more time protecting them and catering to their needs than previous generations. The combination conspires against their independence. "Both of us were more independent at their ages," said Kate, as we watched her fifteen-year-old play soccer. She mentioned her oldest, still in college. "Nina, she's always coming back to us for input about almost everything she does, which I *never* did with my mother."

"Yeah, I didn't check with them so much," agreed Lee.

"It's not even about checking in," said Kate. "It's about checking *what to do.*"

At the same time, adolescents are spending more time with their own cohort than at any other time in the last three centuries, and they're doing so at a moment of furious technological change and mass media influence, which means they're socialized—and socializing—in ways that many parents still find mysterious, even if they too use Facebook (or "Myface," as Hillary Clinton said in a speech at Rutgers University, unintentionally summing up the problem). When I asked Gayle, Mae's mother, what the single hardest thing was for her about raising teenagers, she answered immediately: "Not knowing, or not *really* knowing, what they do." She recalled a time when her youngest, then in ninth grade, told her that she was staying at a friend's house for the night. Her daughter called twice from her cell phone to check in and report that everything was fine. The next morning, Gayle got a call from a youth officer at the Port Authority, wanting to confirm that she'd given her daughter permission to spend the night in New Jersey.

She had not. But a bunch of her daughter's friends had wanted to go, and the cell phone made it easy to lie.

technology and transparency (the xbox factor)

Since the dawn of teen culture, adolescents have led lives apart, but recent advances in technology have given them a whole new mode of asserting their independence and prospecting on their own. "And it's *freaking people out,*" says Clay Shirky, a new-media philosopher at New York University. "Because anything that *you* grew up with that you thought was normal, and your parents grew up with and thought was normal, doesn't just seem like it was normal for two generations. It seems like God wrote it down. Like it was in Leviticus."

Most media revolutions tend to bring with them a squall of public fretting. Social scientists in the 1920s thought movies "fueled cravings for an easy life and wild parties and contributed significantly to juvenile delinquency," according to Mintz. The reaction to comic books a few decades later was arguably worse. In 1954, psychiatrist Fredric Wertham, author of a best-selling polemic against crime comics, told members of the Judiciary Committee: "If it were my task, Mr. Chairman, to teach children delinquency, to tell them how to rape and seduce girls, how to hurt people, how to break into stores, how to cheat, how to forge, how to do any known crime—if it were my task to do that, I would have to enlist the crime comic book industry."

Perhaps the key difference today is that technological change is happening so rapidly it's hard for parents to keep up—whereas their children, whose brains are still plastic and amenable, can adjust to these high-velocity changes in real time. Such adaptations translate into genuine differences of sensibility, which can utterly confound the dynamics of parents and teenagers, even if both sides have the best intentions at heart.

Teenagers, for instance, now have a different sense of time—and therefore planning etiquette—than we do. Fiona, another friend in Deirdre's extended circle, explained it this way: "If my daughter says, 'Oh, tomorrow I want to meet my friends downtown,' and I say, 'What time are you going to go?' *she just won't know.* The plan hasn't occurred yet. Their lives are much more fluid."

I asked why this was a problem.

"Because I want to plan!" she exclaimed. "And they don't know what time the movie is. Then they might meet for pizza, but they don't know—they all have cell phones, they can see at the moment, it'll depend. . . ."

Adults and teenagers may both carry cell phones, but teens use them as tracking devices, the way NASA once followed the space shuttle on a grid. Adolescents are always texting and monitoring each other's whereabouts; they always have an ambient sense of where their peers are. This makes the need for them to plan things (never a teen strong suit to begin with) far less urgent; they can just make stuff up as they go. But their parents—who just happen to be the people responsible for their lives and safety—still have quaint notions about time, observing schedules and verbal agreements, and other concrete, articulated measures. "For older generations," says Mimi Ito, the social media anthropologist, "you explicitly have to open a communication channel—usually with a phone call—in order to meet up face to face. And teens don't do that. Their *default* is that they're always connected. They always have their cells, which means lateness doesn't matter as much anymore."

This sensibility, this whole way of thinking about time and social interaction, is a source of bewilderment and frustration to parents trying to accommodate their adolescents' needs for independence. They want to facilitate their children's ability to have separate and fulfilling lives. But they can't help feeling jerked around in the process. ("We may or may not be going out for pizza later, and we may be at Sam's, but we may be at Jack's.")

The same time management issue rears its head, adds Ito, when kids play video games. These games are "really accessible," she says—meaning on any device—"and they don't tend to have beginnings and ends." When parents spend forever trying to get their kids to stop playing video games and come down to dinner, they're trying to impose artificial boundaries in time where no natural ones exist.

BUT PERHAPS MOST PROBLEMATIC and confusing to parents is the inverted power structure created by their children's technological fluency. The fourteen-year-old becomes the de facto chief technology officer of the house, with parents coming to him or her to ask where Pandora is on their new television or how to close all the windows on their iPhones. Mothers and fathers describe feeling powerless in the face of the new devices in their midst—including the ones they're trying to regulate.

"Not only that," says Shirky, "but parents live in a society that gives them a sense that *their kids* have to give *them* permission to do things. Like asking your child to friend you on Facebook." Everyone talked about this at Deirdre's house. Some women were friends with their kids, some weren't, and some had limited access. Beth told the group she'd been friended and unfriended several times by her son, though her daughter always gave her full visibility into her life. Samantha said her daughter first told her she had a no-friending-adults policy; then it turned out that she was friends with Samantha's first cousin, also in her fifties. Deirdre kept tabs on her kids through her husband, who was on Facebook, though she wasn't. "Friending your child—that's an anxiety-producing activity, no matter what the answer is," says Shirky. If your child says yes, you may see things you hadn't expected; if the answer is no, your feelings are hurt and you're forever wondering whether dubious posts are accumulating on your child's page. "And *that* problem," says Shirky, "is new."

Before social media, both the telephone and the television were semipublic utilities in the house. Even if children locked themselves in their bedrooms with the family telephone, their parents knew they were in there, talking to someone; with a couple of artful questions, they could even find out who without appearing too meddling. The same went for the television: even if parents hated the shows their kids were watching, they could know, simply by breezing through the den, what those shows were, when they would be over, and whether another one was starting.

"What's interesting about cell phones and Facebook," says Darling, "is that there's no way to passively monitor them."

And that's the crucial word, as far as Darling is concerned: "passively." "You have to accept," she says, "that looking at these things is an invasion of privacy. It's *active*. You have to do something that's nosy, and that feels like spying."

For some parents, that's a tough line to cross. It means giving themselves license to snoop when they know (and remember, having once been teenagers too) the value of privacy. In some ways, they probably have more sacred notions about privacy than their children do. Parents have always snooped, of course. There have always been mothers leafing through their daughters' diaries and fathers poking around their sons' rooms for hidden cigarettes. But snooping now feels obligatory, rather than desultory, and therefore extra-intrusive; it's also *work*. There's usually more than one device or platform to consider monitoring (Facebook, Tumblr, Flickr, phone texts, phone photos, Twitter feeds, Xbox hours), and almost all require a touch of savvy to do it. The women at Deirdre's went around and around about this question, debating the ethics of surveillance (or "creeping," as kids call it). One woman said she promised her daughter that she'd never read her Facebook page and stuck to it; another said that she never made such a promise and spied all the time. But it wasn't until Deirdre spoke that a larger emotional truth seemed to crystallize for all of them: the challenge, they realized, wasn't just about giving themselves license to snoop, but about accepting what they'd find.

"One time," said Deirdre, "I got all whipped up about something I found in my spying. And my husband was like, 'Deirdre, maybe you don't need to be looking. It's just upsetting *you.*'"

"Exactly," said Beth. "Sometimes I would rather not know."

Because knowing means running the risk of seeing difficult stuff. Pictures of your drunk kid at a party, because someone posted them on his or her Facebook wall. Naked pictures of your daughter on her cell phone—still unforwarded, as best as you can make out—so why are they there? To send to someone later? Or were they just taken as an experiment, to see what she might look like? (Beth had this experience.) Kate, spying once, saw that her son was making plans to get high. "Do I want to see things I don't want to see," she rhetorically asked, "and try to deal with things I'm not supposed to know about?"

"While we were growing up," says Shirky, "we were clearly experimenting with liquor, but if we didn't come home reeking of gin, there was some meeting place about it"—meaning a tacit agreement it wouldn't be discussed. "And movie theaters were the culturally approved place for teenagers to experiment with kissing each other. We had some set of bargains the parents understood but *didn't talk about.* And that's all broken now."

It's renegotiating those bargains and figuring out ways to cope with new potential modes of transparency that have everyone improvising and scrambling. In this transitional era, there are no norms. And that makes life for mothers and fathers more confusing. Sometimes the policy of "don't ask, don't tell" is easier than grappling with the possibilities of full disclosure.

But for parents who embrace it, this technology can spell good news too. The online world is so seductive that kids now often commit their indiscretions from the safety of their own bedrooms rather than in the real world, where the physical harms truly lurk. "The electronic stuff makes it easier in some ways," Beth told the group at Deirdre's, "because they're doing their naughty stuff online. They're not *out* like I was."

"I know," agreed Gayle. "I always wanted to be out. But they'll come home if they're bored."

Texting also allows children a fast and easy means to communicate frequently and discreetly with their parents—if they're at a party where everyone's too drunk to drive, for instance. And parents can do the same, keeping tabs on their children at moments when they other wise would have been unreachable, and sending them quick notes to say hello if a phone call seems too imposing.

How teenagers use new technology more or less embodies the paradox of modern adolescence: they are doing things you don't know about, but they're doing them under your very own roof, on a computer you purchased. They're using their cell phones to lie about where they're spending the night, and they're using them to text you from college to tell you about their new roommates. (According to Barbara Hofer, a Middlebury psychology professor, first-semester college freshmen are in touch with their parents 10.4 times per week.) The technology of today intensifies an adolescent's dual existence. They are hyperconnected to their families. But they also lead lives quite separate and apart.

whose excesses?

We're still talking about Wesley's egging. I ask Samantha how she reacted to the episode that night.

"I had a freak-out," she answers.

"To this day," says Wesley, staring straight ahead. "Still."

What does her version of a freak-out look like?

"Me screaming like a hysterical person in this kitchen."

Was it worsened by feelings of self-doubt? I ask. Did she wonder if she'd done anything wrong?

"No," she answers. "I was blaming the other kid. I don't think Wesley would have thought of egging on his own. Maybe I'm wrong, but . . ."

Wesley cuts her off. "I initiated it completely." He looks her dead in the eye.

"Interesting." Samantha keeps her cool this time. What's so intriguing is where her mind goes. She looks over at me. "You know, I did feel a *little* responsible," she says, "because we used to keep acorns in the car, and when someone did something stupid, we used to toss an acorn. And I used to wish I had eggs, but I never *threw* an egg at someone." She looks uncertainly over at her son. "So I did think maybe he did get that a little from me."

I doubt he did. But that isn't what matters. What matters is that she can identify with his impulses. She's had them herself.

THE EXCESSES OF ADOLESCENCE are terribly frightening to parents. But the reason, if you read what psychoanalyst Adam Phillips has to say on this subject, is not because these excesses are foreign. Rather, it's because they're so familiar. Parents can identify with them all too well.

Adults are plenty familiar, for instance, with the wish to throw tantrums. (As I noted earlier, kids gave their parents their rock-bottom lowest marks for the ability to control their tempers in Ellen Galinsky's *Ask the Children*.) Adults are plenty familiar with the siren calls of email, video games, Facebook, Internet porn, texting, sexting. Adults are plenty familiar with the urge to drink too much, to have sex in wildly impermissible contexts, and to throw punches at their boss or slam eggs at a nosy neighbor's window. At one point in Deirdre's kitchen, Samantha blurted out what seemed like a non sequitur: "Once I ran away from home for two days."

"I did that all the time," said Beth, reassuringly.

"As an *adult,* Beth!" she answered. "As a *parent.* I left home."

"Adults," Phillips writes, "are not less excessive in their behavior than adolescents. Concentration camps were not run by adolescents; adolescents are not mostly alcoholics or millionaires."

The sole difference, in Phillips's view, is that adults have spent a longer time living with these impulses and therefore, with any luck, have learned to tolerate them rather than act on them. Alas, that's what adulthood is supposed to be about: "an overcoming" or (better yet) "a disciplining of a developmentally appropriate insanity." Adolescents are just a reminder that this insanity remains holed up somewhere within us, waiting to wriggle to the surface.

We may envy that insanity as much as we fear it. But as adults, the most we're allowed to do is sublimate our chaotic feelings. We're forbidden to act on them directly. "Adolescents, and their parents who were once adolescents, are simply experiencing two kinds of helplessness," Phillips observes. "The helplessness born of experience, and the helplessness born of lack of experience."

Mintz notes that although grown-ups like to treat teen problems as if they're strange and distinct, they in fact rise and fall in tandem with adult problems. If you survey the data of the last twenty-five years of the twentieth century, trends in smoking, drinking, drug use, out-of-wedlock births, and violence tended to follow the same-shaped curve for both groups. It's just that adults like to project their anxieties downward, on a generation they believe they can control.

Or not. Beth's ex-husband, Michael, is a recovering addict. He started abusing drugs and alcohol in his late teens, and he didn't stop until his late twenties. To this day—he's fifty now, a project and installation manager of security systems, and remarried to a lawyer—he suspects that he spends too much time behaving like a kid around his teenage son and stepson. "I probably stoop too much down to their level," he told me over coffee, "messing around with them too much."

He was perfectly upbeat as he said this. I got the sense that one of the not-so-secret pleasures of having teenage boys around, at least for him, was the chance to play roller hockey and indulge in all-around horseplay. But at the same time, he was sober now, in all senses of the word, and knew what a mature adult life required. "You have to build a family," he solemnly told me, "out of sand and stone."

memories, dreams, and reflections

Perhaps this is what's so powerful about adolescence for parents: we're now contemplating ourselves as much as we're contemplating our own children. Toddlers and elementary school children may cause us to take stock of our choices, and they may even awaken feelings of regret. But it's adolescents, usually, who stir up our most self-critical feelings. It's adolescents who make us wonder who we'll be and what we'll do with ourselves once they don't need us. It's adolescents who reflect back at us, in proto-adult form, the sum total of our parenting decisions and make us wonder whether we've done things right (whereas young children are still unformed, still works in progress; there's still time to change course if need be).

As part of his study of the parents of adolescents, Laurence Steinberg asked his participants to fill out a "mid-life rumination scale," which included this item: "I find myself wishing I had the opportunity to start afresh and do things over, knowing what I do now." Nearly two-thirds of the women reported frequently feeling this way. So did more than half the men.

When he wrote up the results for *Crossing Paths,* Steinberg made a crucial distinction about this question. He noted that the survey item didn't ask participants whether they wanted to be *teenagers* again. That's the clichéd wisdom—that what adults truly crave in midlife is the raucousness and freedom of their youth (thus the clichés about men purchasing red sports cars and women running off with their tennis instructors). What Steinberg realized, in follow-up interviews with his subjects, was that they didn't want a second adolescence at all. "What they want," Steinberg writes, "is a second *adulthood* [emphasis mine]." Their children's adolescence, he found, was often cause for extensive inventory-taking, if not a full-scale review of their life choices. "Filled with misgivings about their choice of career, spouse, or lifestyle," he writes, "they want a chance at another life."

This inventory-taking is precisely what Gayle does when I sit with her in her sunny kitchen on a Sunday morning as her three adolescent daughters, ages fourteen, seventeen, and twenty, slowly start to stir. She relates a counterfactual history of herself. But she doesn't rewind the tape to the very beginning. She rewinds it to the moment when she left home. "If I had chosen to spend more time on studying in high school," she says, "I could have gone to a different college, finished sooner, and maybe been in a different career earlier, which would have sent me on a different path. And maybe in the scheme of my life, that would have been a better choice."

Gayle's choice was to be a stay-at-home mother. When she made her decision, it made perfect emotional sense. "I quit working because I couldn't stand being away from my children," she says as her girls yo-yo in and out of the kitchen. "To be away for an hour, to go to the bank, just *hurt* me." She never once thought less of her friends who continued to work and found alternative child care arrangements. It just wasn't something she could get motivated to do herself. "And now I think, *What kind of role model was I?*" she asks. "I have three *girls* and I *quit my job?* I went to college and grad school!" She shakes her head. "If they'd been boys, maybe that wouldn't have bothered me so much."

Her kids have noticed her choices. "I know for sure Lena"—her middle child, seventeen—"has said to me, 'Why didn't you work more?' And I say, 'I wanted to be with you.' "

And how does Lena respond to that?

"She's polite. She's not going to say, 'I don't really need you.' But I stayed home too long. I know that now."

When her children hit adolescence, Gayle tried to reverse course. For a while now, she's been looking for work as a public school teacher, the field in which she was originally trained. But trying to find a job in a sector that's suffering terrible cutbacks—at the age of fifty-three, no less—is not easy, and the process has not exactly boosted her morale. "If you look at any school review," she says, "you see, 'The teachers are young and energetic.' Which sounds great. But for me, it's a little blow. I think I'm energetic. But not young."

Over the years, Gayle has also been forced to reckon with the financial consequences of her decision to stay at home. She recalls one of the road trips she took with Mae in her junior year, touring some of the schools in New York State's university system. They quarreled bitterly. Mae thought the quality of some of them was so low that it was a waste of time to apply. "And I was saying, you'd *better*," says Gayle. Those colleges were what she and her husband, who owns a small mail-order business, could realistically afford.

As Mae was growing up, Gayle conveyed to her the idea that she could go to any college she wanted, so long as she worked hard enough. It was a useful illusion, one spun primarily out of love—to make Mae feel secure, to make her feel optimistic, to make her feel confident and powerful and motivated in a world that is in fact sometimes scary and hard to navigate. "You raise children to think the world of possibilities is theirs," says Gayle. "And we somehow think, 'Oh, we'll make enough money,' or, 'Oh, they'll get in on a soccer scholarship.' And then, all of a sudden, they're eighteen and it's like, 'Oh no, you can't go to college there.'"

On that road trip, Mae called her mother's bluff. She assessed with a gimlet eye the limitations of the world around her and declared she didn't like them. That was when Gayle realized that this spell she'd cast, this story she'd so lovingly told, was perhaps as much for her own benefit as it was for Mae's. "We," she tells me, "had been living in that dream world too."

ERIK ERIKSON, ONE OF the most innovative psychoanalysts of the twentieth century, wrote about these moments of existential review in his work on the human life cycle. He famously argued that all of us go through eight stages of development, each marked by a specific conflict. That he thought to extend his model to include adult life—to even conceive of adulthood as a series of hiccups and pivots rather than an unbroken forward march—has given his theory remarkable staying power, and the stages of adulthood he identified are very relatable. In early adulthood, he argues, we must learn how to love rather than vanish in a mist of narcissism and self-protection. In mid-adulthood, he says, we must figure out how to lead productive lives and leave something behind for future generations rather than succumb to inertia ("generativity versus stagnation," he calls it). And following that, the challenge becomes learning how to make peace with the experiences we've had and the various choices we've made rather than capitulate to bitterness ("integrity versus despair and disgust").

Some modern researchers believe that these adult stages are overstated, even fanciful inventions. But the parents of adolescents often describe them to the letter. They talk, as Gayle did, about hoping to fight off stagnation going forward, though they face a diminished range of career options. And they talk, as Gayle did, about looking backward and integrating the choices they've made into a narrative they can live with. In Erikson's words: "It is the acceptance of one's one and only life cycle and of the people who have become significant to it as something that had to be and that, by necessity, permitted of no substitutions."

Women may be especially susceptible to these moments of self-reckoning. According to the 2010 Current Population Survey, 22 percent of all parents of twelve- to seventeen-year-olds are now fifty or over, and 46 percent of them are forty-five and over. What this means, practically speaking, is that a substantial number of today's mothers of adolescents are either in perimenopause—experiencing hot flashes, sleep disruptions, and changes in sexual desire—or in menopause itself. Many women pass through this stage with little turmoil, just as many adolescents pass through puberty with little ado. But others struggle with melancholy and irritability, seeing in their condition a mirror image of their teenagers, whose fertile years are just beginning. (A pair of well-designed studies from 2006 found that the risk of depression during perimenopause either doubles or quadruples, depending on whose numbers you consult.)

Gayle, happily, says that she loves seeing her daughters bloom. It's a form of compensation rather than a rebuke. "I get a lot of pleasure watching them change from girls to women," she says. "Their sexuality doesn't bother me at all. Being closer to death does. I just think, *God, I'm old*."

She isn't old. She's just fifty-three. Still, she gestures toward the living room, where Eve is sitting. Gayle had her at thirty-eight. "I look at her," she says, "and I think, *When I was her age, it was almost forty years ago*."

REGRET SHOWS UP IN all kinds of strange dress. Sometimes it comes out purely as questions about oneself—the career one should have pursued, the lifestyle choices one should have made, the spouse one should or shouldn't have chosen. The mere presence of adolescents in the house, still brimming with potential, their futures still an unclaimed colony ("My girls are about to make choices of their own," Gayle told me), sets off a fantastical reverie of what-ifs.

But sometimes this regret comes in the form of doubts over how we've raised our children—over parenting itself. This regret can be subtle, and not necessarily about things we've consciously chosen to do. Some of the worst pain, in fact, can stem from the things we *failed* to do, or the errors we made that our kids have seen, or the rotten habits we failed to conceal and our kids have now made their own or decided very aggressively to reject. Children bear witness to some of our most shameful behaviors and worst mistakes. Most parents can tell you with grim precision what they were, and the pain those habits and episodes inflicted.

It doesn't help that adolescents often take a harsh, unnuanced view of their parents' flaws and mistakes. It becomes the device they use to push their parents away—to distinguish themselves ("I will not become you"), or "individuate," as a psychologist would say. They know how to transform their worldview into a weapon, into observations very precisely tailored to hurt. One of Calliope's go-to phrases that she sometimes uses to hurt her mother, Samantha says, is to tell her that she has turned into her own mother—and the last person Samantha would like to be is her mother, whom she considers terribly cold and neglectful. During one argument, Calliope even called Samantha by her mother's name. "I knew that was hitting below the belt," Calliope says.

The most profound parenting regrets I hear, though, come from Michael, Carl's father and Beth's ex-husband. When we sit down to speak, I can tell that Michael isn't the kind of guy who *really* regrets. He considers himself a lucky man, all in all, and believes that his kids will ultimately sense his good intentions. But when his kids are giving him grief, his mind loops back to the days when he and his ex-wife were hammering out the terms of their divorce and he failed to press her for joint custody. "It would have been a fight in court, it would have continued the circle of fighting, it didn't make sense," he says. "But I can still regret it."

He knows he's paid a price for it, especially with his older daughter, Sarah. "My relationship with her has always been fractured," he says. "We've never been totally comfortable together." He recalls the day a couple of years ago when she graduated from high school. There she was, radiant and full-grown, a young woman who'd gone to a great public high school and gotten a near-full scholarship to an even finer private college—his own daughter! he never even *went* to college—yet he felt awkward, estranged, grounded amid a swarm of airborne balloons. "The graduation's over," he remembers, "and it's like, 'Okay, where's everybody going?' And the answer is, 'Everyone goes to Mom's.'" He gestures away from himself, to an imaginary other location—*Mom's.* "So I'm over here"—he points to himself—"all dressed up, thinking, *I can probably invite myself, but it's not going to happen.* It's almost a feeling of . . . helplessness. She's my daughter too." He can't talk about it without detaching a bit, switching from the first to the second person. "It's like you're not part of this," he says. He reflects on this sentiment and then owns up to it. "That's how I felt. Like I wasn't part of this."

And when his son, Carl, is feeling cruel, or angry, or even merely defensive, "he'll say, 'Sarah doesn't want to see you; she doesn't like you,'" says Michael. "If he wants to throw me off, that's pretty much where he starts." Michael means it when he says, "starts." These assaults sometimes escalate in unbearable ways. "It's like having an argument with one of your friends who's being vicious," says Michael. "And then, you sit around and think, *Does what he said apply? Does it not apply?*" And in some cases, Michael decides, it does. "He's made me cry before," Michael admits.

outcomes

Gayle's middle and youngest daughters, fourteen and seventeen, are easygoing and placid. They may have their moments of adolescent testiness, but they usually speak with affection when they're around their mother, and this morning they move quietly through the kitchen, taking up their morning chores without complaint.

And then there's Mae, who also spends time with us in the kitchen. She's a lovely-looking girl, a long-stem rose like her sisters, but the air around her vibrates; there's a sense of vigilance about her, a worry, as if she already knows the road ahead will be hard. She's my kind of girl, truth be told. Anyone who has early intimations of life's difficulties is my kind of kid. I was that kind of adolescent myself.

"Am I peeling?" She turns around and shows her mom her back. She, like both sisters, is wearing a tank top; she also sports a discreet stud in her nose.

Her mother answers that she isn't.

Mae was always different. Gayle could see she was an anxious kid, even at five. In fifth grade, when cliques started to form, Mae was having trouble with her best friend, Calliope, Samantha's kid, and there was little that Gayle could do to ease her anguish. "Mae would have this thing, where Calliope was mad at her; she didn't know why," recalls Gayle. "So she'd follow her around and say, 'What did I do?' And I'd have to say, 'Do not do that.' " Just the memory of it makes her cringe, both reexperiencing her daughter's misery and knowing she had to let her cope with it on her own.

Then, in eighth grade, Mae started cutting herself. Gayle didn't know anyone else whose child struggled with the same problem, though she'd heard and read plenty about it; this was an enlightened generation of parents in an enlightened community. So Gayle did what she could: she found her daughter a therapist to talk to, and she learned to listen and, when appropriate, to offer advice. And her daughter got better. Looking at her now, you see a pretty, extraordinarily thoughtful kid who's gotten herself almost a full ride through a great university.

But looking at Mae, one also sees fairly clearly what Adam Phillips means in *On Balance* when he says that happiness is an unfair thing to ask of a child. The expectation casts children "as antidepressants," he notes, and renders parents "more dependent on their children than their children are on them."

Just as important, Mae is a good example of why producing happy children may not be fair to ask of *parents*. It's a beautiful goal—one I've readily admitted to having myself—but as Dr. Spock points out, raising happy children is an elusive aim compared to the more concrete aims of parenting in the past: creating competent children in certain kinds of work; and creating morally responsible citizens who will fulfill a prescribed set of community obligations.

The fact is, those bygone goals are probably more constructive—and achievable. Not all children will grow up to be happy, in spite of their parents' most valiant efforts, and all children are unhappy somewhere along the way, no matter how warmly they're nurtured or how stoutly they're protected. There are, in the end, crude limits to how much parents can do to shield their children from the sharper and less forgiving parts of life—which, as adolescents, they stumble on far more regularly. "For a child growing up," Phillips writes, "life is by definition full of surprises; the adult tries to keep these as surprises, rather than as traumas, through a devoted attentiveness. But sane parenting always involves a growing sense of how little, as well as how much, one can protect one's child from; of just how little a life can be programmed."

To this day, Mae feels things more deeply than her peers. And Gayle, possessed of a calm, unfussy Midwesterner's temperament, does not blame herself for this, as perhaps another parent might. She has as much compassion for herself as any mother can expect to have, knowing she's done all she can humanly do for her daughter. "It's not that I feel inadequate as a mother," she tells me. "I feel the inadequacy as a human to solve *any* other human's problems. You can only help another person so much." But that doesn't mean it's easy. When I ask if she's learned how to better cope with having an anxious child over the years, she answers immediately: "No."

AND YET HOW PROUD Gayle is of Mae! How amazed, how full of admiration! There came a moment when I mentioned Erik Erikson out loud, wondering if Gayle had ever heard of him. Gayle said he sounded familiar, but no, not really. Mae, who'd been silently lingering at the kitchen counter, left the kitchen, went upstairs, and retrieved a copy of a book by Erikson, which she'd been reading for psych class. She plunked it down in front of her mother. Then she quietly left the room.

Gayle smiled at me.

"That's the kind of thing you live for," she said. "You *want* them to be better than you. You want them to be smarter and do more things and know more things." She picked up the book and scanned both its front and back cover. She'd already mentioned to me that she loves Mae's writing, loves her mind. "Gosh. I didn't read this when I was twenty."

And that's just it. In spite of our mistakes, here they are, thoughtful and accomplished human beings, gesticulating with our mannerisms and standing at our height.

Back at Samantha's house, there came a moment when she wondered aloud whether she hadn't focused enough on Wesley when he was small. "I just remember when Calliope was little," she said. "Wesley was always being awakened from a nap and scooped up in a car seat and put someplace. His standards were so much lower, in terms of his demands. He was just happy to have *food.* And I thought, *I wonder if I've done this to him.* But I look at his dad, and *he's* like that. I don't know how you feel, Wesley. . . ."

She then looked directly at her son—so talented, so perceptive, and Lord, such a pain in the ass sometimes, causing her so much grief. Yet it wasn't a look of desperation to validate her choices. It was a brave thing she did. She seemed genuinely to want to know. He looked back at her, then uncertainly into the middle distance, with his hands cupped over the neck of his guitar. Several seconds ticked by, then several more. It was the only time there was no guitar-playing running in the background of our interview. It was the only time there was no sound at all.

"Start speaking when you're ready," said Samantha. But it wasn't Wesley who needed the extra time. It was she. "I just feel like having kids is the greatest thing I ever did, and I . . ." Her voice caught, and she started to cry. "I'm so proud of them. I love them so much. Last night, I was remembering when Calliope was a baby and being like, *Oh my God, that's so* gone." Her kids, startled by this frank display of emotion, looked at one another and themselves started to well up. "And then I thought, *Well, someday maybe she'll have a baby too. . . .*" Samantha wiped her nose.

Wesley still said nothing. Calliope, almost never at a loss for words, said nothing either. She put one hand over her mouth. With the other, she laced her mother's fingers in her own.

chapter six

joy

But I am telling only half the truth. Maybe only a quarter of it. The rest of the truth is that I was unable to bear loving my children so much. Loving left me weak, skinless. Ideally I would have liked Katherine and Margaret sewn to my armpits, secured to me. Or, better yet, kicking and turning in my stomach, where I could keep them safe forever. —Mary Cantwell, *Manhattan, When I Was Young* (1995)

THROUGHOUT THIS BOOK, I'VE tried to look at how children affect their parents at each stage of their development. To do so, I've examined inflection points and sources of tension, hoping to identify which ones are universal and which are unique to this moment in time. In chapter 1, I tried to explain how children can compromise the autonomy we've grown accustomed to, making it harder to sleep, harder to find flow, harder to manage the boundaries between our work and home selves. In chapter 2, I talked about the still-stalled revolution at home, and how a lack of consensus about domestic divisions of labor puts added strains on a marriage once a child comes along—strains only made worse by declines in social support. In chapter 4, I talked about concerted cultivation and the pressures on parents, since World War

II, to intensively parent, pressures abetted by the soaring cost of living; ambivalence about women in the workforce; the rapid churn of technology; increased fears about child safety; and most of all, the economic inutility of children in the modern age. And in chapter 5, I examined the effect of the prolonged, protected childhood—as well as the culture-wide pressure to produce "happy, well-adjusted children"—on the parents of adolescents.

But each chapter (save for chapter 3, about the joys of young children) has had a built-in bias. Each has focused on parenthood as we actively pursue it day to day, rather than on what parenthood actually *means* to us, or how the overall experience of being a parent is swept into our self-image.

There's a reason I've tilted in this direction. To raise a child requires galactic effort, and since the advent of modern childhood—since the moment "parenting" became a popular verb, especially—there has been an even greater emphasis on child-rearing as a high-performance, perfectible pursuit. To repeat the words of William Doherty, raising children is "a high-cost/high-reward activity." Chapter by chapter, I've tried to document those costs.

But those costs don't mean that children can't provide us with moment-to-moment pleasure. Robin Simon, a Wake Forest University sociologist whose findings on the relationship between parenting and happiness rank among the most negative in the social science canon, said this to me outright, over coffee: "There's really *fun* stuff about raising kids." That's right. *Fun.* She sees no contradiction between what she's found in her research and this idea. She mentioned her nineteen-year-old son. At that moment, he was going through a karate film phase. "It's *fun* watching really bad films with him," she said. "It's *fun* to hear his ideas about things and to see him express his interests." It's just that the fun parts of raising a kid—whether it's singing at the top of your lungs or buying your daughter a dress, coaching a soccer game or staying in and baking banana bread—can be overwhelmed by the strains and moment-to-moment chores of the job.

But mothering and fathering aren't just things we do. Being a mother or being a father is *who we are.* "When I think of the word 'parenting,'" writes Nancy Darling, "I think of asking my kids to set the table, getting them to do their homework, or getting my youngest to practice his violin." She notes that this is hard work. More to the point, it is "probably the least pleasurable part of my interactions with them."

So what gives Darling pleasure?

> Hanging out watching videos, having tea, having them spontaneously come over and hug me, being amazed at how good they are at doing things and how little push they need to do what they're supposed to, and the quiet wonder of just being with them watching them GROW. . . . Listening to my youngest practice a violin etude last night—one he had bored himself silly with all summer—I was just awestruck that this kid who has terrible handwriting, who likes fencing with sticks in the backyard and who will do anything to start a water fight—could make such truly beautiful music.

And what those pleasures have in common, she realizes, is that they're passive. "They involve just sitting back," she writes, "and enjoying my kids being themselves." They don't show up as readily on surveys and questionnaires. "If you asked me about how I felt about parenting," she concludes, "none of those pleasures would be assessed." How it feels to *be* a parent and how it feels to *do* the quotidian and often arduous task of parenting are two very separate things. "Being a parent" is much more difficult for social science to anatomize.

joy

We live in an age where we're told that striving for happiness is paramount. Our right to pursue it is enshrined in our nation's founding document; it's the subject of innumerable self-help books and television shows. Happiness is the focus of a burgeoning field in academia called positive psychology, which studies what makes the good life and all-around flourishing possible. (For a while, positive psychology was the most popular course among undergraduates at Harvard.) Happiness, we are told, is achievable. When we're surrounded by so much material prosperity, as we are today, it is our prerogative—our due, even our *destiny*—to attain it.

But "happiness," as so many parents have said to me, is a big and hopelessly imprecise word. One of the women at an ECFE class, a grandmother named Marilyn who happened to be visiting that day, put it as bluntly as I've ever heard it when she asked: "Shouldn't we make a distinction between happiness and joy?" To which all the people in the class agreed that, yes, we probably should. "It seems to me," she said, "that happiness is more superficial. I don't really know how others feel, but for me, having my kids brought me this deep sense that I've done something worthwhile in my life. . . ." And then she started to cry. "Because when all is said and done, and I ask, *What was my life about?*—now I know."

Meaning, joy, and purpose come from a great variety of sources, not just children. But what's important here is Marilyn's more basic observation: a single word, "happiness," often cannot fully capture these feelings—or countless other emotions that make us feel transcendently human. The awed, otherworldly feeling you get when your infant looks directly into your eyes for the first time is different from the sense of pride you experience when that same kid, years later, lands a perfect double axel, which in turn is different from the sensation of warmth and belonging that consumes you when your widely dispersed family gathers for Thanksgiving. You can try to quantify each of these feelings with a number, certainly, and I don't underestimate the value of such attempts, for the sake of figuring out ways to get more of them. But in the end, a number is just that—a digit, plotted on a graph. It may reflect the degree to which we feel something, but it entirely lacks dimensionality. These feelings do not strike me as the same *in kind*. Some, like joy, can hurt almost as much as they elate; others, like duty, run silently in the background, perhaps making our day-to-day lives more difficult, but making our overall lives more worthwhile and more consonant with our values.

"Few of the experiences of happiness that are conveyed in autobiographical writings and literature," writes Sissela Bok, the Harvard philosopher, in *Exploring Happiness*, "can be fully measured by psychological or neuroscientific research." Nor, she writes, can "contemporary measures of happiness convey most of the philosophical and religious claims about the nature of happiness or about the role it plays in human lives."

Indeed, one could argue that the whole experience of being a parent exposes the superficiality of our preoccupation with happiness, which usually takes the form of pursuing pleasure or finding our bliss. Raising children makes us reassess this obsession and perhaps redefine (or at least broaden) our fundamental ideas about what happiness *is*. The very things Americans are told almost daily to aspire to may in fact be misguided. (There's that line in *Raiders of the Lost Ark* that Sallah and Indiana Jones utter in unison: "They're digging in the wrong place.") As we muddle our way through the parenting years—trying to make sense of our new role in the age of the priceless child, trying to execute that role in a culture that provides so little support for working and nonworking parents alike—it is very worth asking: what *are* we digging for, and what have we found?

LET'S START WITH JOY. It wasn't just Marilyn who used this word to describe her experience. Practically all parents do. And perhaps no person alive has thought through the idea of joy with more thoroughness and more care than George Vaillant.

Vaillant is a psychiatrist by training, a poet-philosopher by temperament, and, from the point of view of the history books, the decades-long steward of the Grant Study, one of the most ambitious longitudinal surveys in social science. Since 1939, the Grant Study has followed the same group of Harvard alumni, all sophomores when the project began, collecting data about every aspect of their lives (and, by now, their deaths). Not surprisingly, then, Vaillant tends to take the long view of things rather than focus on moment-to-moment happiness. "Their lives were too human for science," he once wrote of the Grant Study men, "too beautiful for numbers, too sad for diagnosis and too immortal for bound journals."

When I first meet Vaillant in Boston, he's dressed in a cheerful blue sweater with holes in it, which seems of a piece with his cheerful, slightly abstracted demeanor. He has dense eyebrows, lively eyes, and an unusually erect bearing for a fellow of seventy-seven. "Your generation can't imagine a world without attachment," he tells me as we settle down to chat. "But fancy: before, when behavioral scientists wrote about love, it was all *sex*." He's primarily talking about Freud and Skinner, who couldn't begin to examine the love between parent and child without seeing eroticism. "They couldn't *conceptualize* attachment."

Yet that's exactly where joy comes from, according to Vaillant: attachment. In his book *Spiritual Evolution,* he writes, "Joy is connection," simple as that. Joy is very different from the kind of pleasure one gets from pursuing excitement or satisfying a drive. Those pleasures tend to be intense and ephemeral. "It's how Freud saw sex," says Vaillant. "A full prostate, and releasing it is glorious."

Vaillant doesn't want to shortchange such pleasures. He recognizes that we're wired for them. They're *fun.* But also solitary. They're very different from joy, which is almost impossible to experience alone. "It's the difference between watching *Emmanuelle*"—a famous erotic French film from the 1970s—"and watching Thanksgiving dinner cooked in Grandmother's kitchen," he says. "They're both forms of pleasure." But the first turns the individual inward, while the second turns the individual outward, toward others. The second is what intrigues Vaillant. "It's watching your grandmother," he says, "who's too fat, and your mother, who's got too many improving ideas, and your little brother, who's chasing you." That familiarity, that sense of bondedness, and "those kitchen smells," he says, "are all about Thanksgiving connection." Joy is about being warm, not hot. In *Spiritual Evolution,* he offers this lovely maxim: "Excitement, sexual ecstasy, and happiness all speed up the heart; joy and cuddling slow the heart."

Some of the most poignant testimony I heard from parents was about this need for connection. As Angelique, the mother of four in Missouri City, was describing her thirteen-year-old son, who'd lately taken up football, I asked her what made that age magical. "When he stands in front of me and wants a hug," she told me. "They still *need* hugs, thirteen-year-old kids." Leslie, who lived in the neighboring town of Sugar Land, told me something similar about her ten-year-old: "He'll say, 'Can I go to so-and-so's house?'" She pantomimed a nod, a wave of the hand—*Go, shoo.* "And he'll be halfway out the door," she continued, "and then he'll say, 'Oh, I forgot something,' and he'll turn around and run into the kitchen and give me a hug."

They seem so jaunty and independent, playing on their Wiis or hauling off to practice in their oversized football gear. But all they want, and what they need most of all, is you. And you, them.

But connections, no matter how strong, are still made of a thousand gossamer threads. If that's what joy is—connection—then to fully experience it requires something terrifying as well as exalting: opening oneself up to the possibility of loss. That's what Vaillant realized about joy. It makes us more vulnerable, in its way, than sadness. He's fond of quoting William Blake's *Auguries of Innocence:* "Joy and woe are woven fine." You can't have joy without the prospect of mourning, and to some people this makes joy a difficult feeling to bear.

Especially in parenthood, where loss is inevitable, built into the very paradox of raising children; we pour love into them so that they'll one day grow strong enough to leave us. Even when our children are still young and defenseless, we feel intimations of their departure. We find ourselves staring at them with nostalgia, wistful for the person they're about to no longer be. In *The Philosophical Baby,* Alison Gopnik uses the Japanese phrase *mono no aware* to describe it: "bittersweetness inherent in ephemeral beauty." Joy and loss are part of the inherent contradictions of gift-love. "We feed children in order that they may soon be able to feed themselves; we teach them in order that they may soon not need our teaching," wrote C. S. Lewis. "Thus a heavy task is laid upon Gift-love. It must work towards its own abdication."

For some parents, fear and joy are even more deeply intertwined. In a 2010 lecture that has since been seen by hundreds of thousands, the University of Houston's Brené Brown began with the following challenge:

> Christmas eve, beautiful night, light snowfall, young family of four in the car on the way to grandma's house for dinner. They're listening to the radio station, the one that starts playing the Christmas music, like, right at Halloween. "Jingle Bells" comes on. The kids in the back seat go crazy. Everyone breaks into song. The camera pans in on the faces of the kids, mom, dad. What happens next?

She told the audience that the most common answer is "car crash." In fact, 60 percent of all people who respond to this question say "car crash." (Another 10 to 15 percent have "equally fatalistic answers," said Brown, "but more creative.") She acknowledged that this reflex could simply be a demonstration of how well we've all internalized Hollywood's lurid imaginings. But she suspects it's more than that. The fact is, scores of parents tell her the same thing, describing real-life situations. She gave a typical example: "I'm looking at my children, and they're sleeping, and I'm right on the verge of bliss, and I picture something horrible happening."

Brown calls this feeling "foreboding joy." Almost all parents have known some form of it. All parents are hostages to fate. Their hearts, as the late Christopher Hitchens wrote, are "running around inside someone else's body." So much vulnerability can be agonizing. But how else can parents experience ecstasy? How else can they know awe? These feelings are the price mothers and fathers pay for elation, and for fathomless connection. "Joy," writes Vaillant, "is grief inside out."

ALL OF WHICH TAKES me back to Sharon, the Minnesota grandmother raising her grandson, Cam. She has faced the unimaginable, burying not just one child but two. In the case of Michelle, Cam's mother, Sharon at least got to see her child reach adulthood, even motherhood. But with her firstborn, Mike, Sharon never got that privilege. He died in 1985, at the age of sixteen.

Sharon and her family were living in Tucson at the time. Michelle was angry and had an IQ of 75, which posed one kind of problem; Mike was angry and had an IQ of 185, which posed quite another. His brilliance, rage, and loneliness flared early. At four, he spent a lot of time on his own, memorizing the spellings of long words ("Constantinople," the perennial "antidisestablishmentarianism"). "And he was always making very weird little jokes," said Sharon, "that none of his peers could understand." Briefly, in elementary school, Mike went through an outgoing phase. "But then he moved it inward, like it was up to *him* to save the world," said Sharon, "and then he *was* trying to save the world." He'd sit in the park, for example, hoping to catch the men who were beating up the local homeless population. This was in sixth grade. Then Mike went to a "gifted and talented" junior high and found his crowd, kids who played *Dungeons & Dragons,* made up languages, wrote poetry. But their fellowship wasn't enough to subdue his depression, which grew especially severe in high school. He started talking to Sharon about how much he was suffering—how he was invaded sometimes by the impulse to take his own life, which he eventually did.

"He came into my room on a Thursday and said, 'I'm feeling suicidal, I think I should go back to the hospital,'" said Sharon. She told me all this while Cam was napping. "So we called the doctor, and the doctor said, 'No, he needs to stand on his own two feet, and you need to stop intervening for him.'" He added that Mike should be in charge of his own medication from now on. "So he was," said Sharon. "And that's what he chose to do with them." She found him the following morning.

How, I asked her, does she make sense of her son's life now, all these years later?

She didn't answer right away. "When I think about my life with Mike . . ." She trailed off. "I don't know. It's so big." She rummaged around for a starting point and landed on perhaps the most logical one. "From the day he was born, I expected a girl. In those days you didn't know. And it took me a couple of weeks to adjust to the fact that he was a boy. But, I mean, he was beautiful. Blond-haired, blue-eyed. Perfect little body. Size and everything. He was just such a . . ." Again, words deserted her. "He was really the joy of my life." His depression eventually showed, and his anger. "But that was always a piece of a bigger life," she said. "He was funny, he was helpful. Going to the mall when he was twelve, he would *still* hold my hand. Walking around. I don't know. He was a great kid. I was proud of him. I always hoped we could find some way to help him. I don't know how to sum all that up."

I hadn't asked that, exactly. I'd been afraid to be too direct, I think, fearing the question would seem either naive or cruel. But what I'd meant was, did she ever despair about what all of it was *for?*

She considered the question for a while. "I don't think so," she said, after a bit. "I meant to have a child; I had a child. That child had an illness, but he was still a whole person. I raised him. We interacted. I wish he had made a different choice. I wish he were still alive today. But . . ." She only paused briefly here. Her answer was simpler than I'd expected. "Raising Mike was still raising Mike," she said. "I am *still* his mother. The fact that he died at sixteen doesn't change the 'what for' for me, in the same way that it didn't with Michelle dying at thirty-three. I still have their lives. They're still my children."

They are part of her history. They are people she loved, people she nurtured, people she sometimes failed, people she sometimes rescued, people who made her feel the best and worst she has ever felt. "They've still brought a full parenthood to me," she said. "It's not a full happiness. It's not a full sorrow. It's a full parenthood. It's what you have when you have kids."

duty, meaning, and purpose

Almost by definition, we associate children with the future. In the crudest evolutionary sense, that's why we have them: to see ourselves—to see our species—continue.

But there's a difference between viewing our children as a continuation of our own DNA and burdening them with our hopes, which may or may not be met. Those who let go of having too many personal expectations may in fact have a healthier attitude toward child-rearing.

In his memoir *Family Romance,* the English novelist and critic John Lanchester makes a beautiful plea. Specifically, he calls for a revival of the concept of duty. "'Duty,'" he writes, "is one of those words that has more or less vanished from our culture. It—the word, and perhaps the thing as well—exists only in specific ghettos like the armed services." And then he turns, almost instinctively, to the topic of caring for others:

> We often prefer to use "care" or "carer" for people who would once have thought that what they were doing, in, say, looking after incapacitated relatives, was a duty. To call the act of changing someone's soiled underclothing a work of caring can make you feel as if you should be doing it because you want to do it, whereas the idea that you're doing it because it's your duty makes it more impersonal and therefore—to my mind, anyway—a lighter burden. It leaves you free to dislike what you are doing while still feeling that you are doing the right thing in doing it.

Children are not the same as incapacitated relatives. And Lanchester is not saying, at any rate, that caring for others can't be pleasurable, or that it can't be something one very much wants to do. But by removing pleasure from the equation, he alters our expectations—essentially by giving us permission not to have any expectations.

It's a liberating thought in an age when kids are not only planned but aggressively sought, through fertility treatments, adoption, and surrogacy. Having worked so hard to have children, parents may feel it's only natural to expect happiness from the experience. And they'll find happiness, of course, but not necessarily continuously, and not always in the forms they might expect. Those who start with Lanchester's very simple idea—that they will love and they will sacrifice—are probably at a great advantage. Finding pleasure in the idea of duty alone goes a long way. As I noted in chapter 1, freedom in our culture has evolved to mean freedom *from* obligations. But what on earth does that freedom even mean if we don't have something to give it up *for?*

Mihaly Csikszentmihalyi contemplates this idea quite a bit in *Flow.* He brings up Cicero's observation that in order to be free, one has to surrender to a set of laws. In our personal lives, Csikszentmihalyi writes, rules can liberate us even as they bind: "One is freed of the constant pressure of trying to maximize emotional returns."

Jessie told me that she and her husband, Luke, came to this same realization as soon as the rules of their own lives multiplied. "We both became happier," she said, "when William"—their third—"was born. It was the tipping point between having an independent life and diving into parenthood. With one or two, you can *kind of* pretend you still have an independent life. But with three, we accepted our lives as parents. This new reality sets in." Three kids created more definitive rules, more definitive structure. "We actually considered having a fourth," she said.

Sharon, too, seemed to find comfort and structure in making larger commitments. As she told the judge when she adopted Michelle, "Yes, I get it, it's for life." She actively and freely chose to care for Michelle, to make that a part of her daily program. And that's how she still views the experience of having raised Mike and Michelle: it was her life's work, what gave it shape. Everything she did for them, day in and day out, was not yoked to any kind of outcome, either tragic or triumphant. She woke up each morning and took care of them because that's what she'd signed up to do.

One could say that this commitment was part and parcel of Sharon's Catholic faith—of any faith, in fact. ("Set thy heart upon thy work, but never on its reward," Krishna tells his student Arjuna in the Bhagavad Gita.) But it's also part of a parent's creed. We don't care for children because we love them, as Alison Gopnik says. We love them because we care for them.

And this is what Vaillant, too, ultimately told me. He has five children. One of them is autistic. This boy was born at a time when most spectrum disorders didn't have names, and if they did, the pediatric establishment seldom had optimistic things to say about them. I asked Vaillant whether his son made him readjust his expectations about what it meant to be a parent. He knew, after all, that his boy would never lead a life that looked like his or mine. He shook his head.

"I didn't have children because I wanted to have an heir or because I wanted someone to take care of me in my old age," he answered. "I had children for the same reason I like growing grass and I like walking in the mountains. Having children is part of the way I'm wired, and it's easy to go with the flow. I had no expectations."

Perhaps Vaillant is simply a product of his generation. Men of his age don't associate children with self-actualization. They had children because that's what they were supposed to do.

But it may also be that Vaillant developed this dutiful perspective over years of caring for a son with autism. His child may have taught him something about what to expect from parenting, and what not to expect. "Here's what's coming to mind," Vaillant told me a few minutes later, after mulling over my question some more. "And this isn't happiness, but it's certainly love: when my son was six, I had to button his buttons for him." He looked off, and several seconds ticked by. "And tie his shoes." While other six-year-olds were already buttoning their own buttons and tying their own shoes. "And that was a chore," said Vaillant. "But so is, when the grass is long, pushing a lawn mower. And how else are you going to have a lawn?"

ONE OF THE MOST famous thought experiments in modern philosophy is Robert Nozick's "experience machine," which he wrote about in his 1974 *Anarchy, State, and Utopia:*

> Suppose there were an experience machine that could give you any experience you desired. Superduper neuropsychologists could stimulate your brain so that you would think and feel you were writing a great novel, or making a friend, or reading an interesting book. All the time you would be floating in a tank, with electrodes attached to your brain. Should you plug into this machine for life, preprogramming your life's experiences?

His response is no. And many people would instinctively agree. We care about much more than getting our kicks. We long for experiences "of profound connection with others," he writes, "of deep understanding of natural phenomena, of love, of being profoundly moved by music or tragedy, or doing something new and innovative." Just as important, we long for esteem and pride, "a self that happiness is a fitting response to." Implicit in Nozick's experiment is the idea that happiness should be a *by-product,* not a goal. Many of the ancient Greeks believed the same. To Aristotle, *eudaimonia* (roughly translated as "flourishing") meant doing something productive. Happiness could only be achieved through exploiting our strengths and our potential. To be happy, one must *do,* not just feel.

Raising children requires a lot of doing. It's a life of clamorous, perpetual forward motion, the very opposite of Nozick's passive experience machine. Not everyone wants children. But for many—especially those of us who don't have the imagination or wherewithal to create meaning in unconventional ways—having children is a way to exploit our potential, to give design and purpose to a life. Robin Simon puts the finest point on it: "Children are a reason to get up in the morning."

Simon isn't just making an informal life observation. She is stating a statistical truth. Parents are much less likely to commit suicide than nonparents, and most sociologists, starting with Émile Durkheim's 1897 book *Le Suicide*, have speculated that is true for just the reason

Simon cites: parents have ties that bind, earthly reasons to keep going.

Durkheim thought a lot about the benefits of social ties. It wasn't just the bonds between parents and children that interested him, but the bonds between adults and larger institutions. Without them, people feel rootless, disoriented; he described their condition as "anomie." Today we think of that word as synonymous with alienation, but that's not what Durkheim was talking about precisely. What he meant was "normlessness" (from the Greek *anomos,* "without law"). It can be very isolating to live in a normless world. In *The Happiness Hypothesis,* Jonathan Haidt describes it this way: "In an anomic society, people can do as they please; but without any clear standards or respected social institutions to enforce those standards, it is harder for people to find things they want to do."

Once people become parents, they often discover they have a much clearer set of standards to obey and a renewed respect for the social institutions designed to enforce them. Listening to new mothers and fathers talk about what they loved most about the transition to parenthood, I was surprised by this simple, commonly recurring theme: they were now connected more strongly to institutions that normalize a life. They suddenly had reasons to go to church, to synagogue, to the local mosque. They suddenly knew all about their neighborhood schools and parks and block associations; they wanted to become involved in parent-teacher organizations and local politics. And the shadow, parallel worlds of their neighbors with kids, which had once run indistinctly in the background, now began to pop in three dimensions. ("It's kind of freeing," a woman named Jen told her ECFE class. "People will come up to me and just start talking. I love having something to talk about with anybody.") Becoming parents gave them a means to relate to others. They could be seated on a train, waiting at the checkout counter, standing on a long line at the voting booth—and odds were, if the person nearest them had a kid, they had a common set of concerns. "The love we feel for children," writes Gopnik in *The Philosophical Baby*, "has a special quality of both particularity and universality."

The idea that children give us structure, purpose, and stronger bonds to the world around us doesn't always show up in social science data. But it can, if you use the right set of instruments. Robin Simon, for instance, has found that parents who have custody of their children are *less* depressed than those who don't. That's a big departure from most other parenting-and-happiness studies, which suggest that single mothers (who more frequently have custody of their kids) are less happy than single fathers. But there's a difference between Simon's study and others: she was measuring *depression*, and depression surveys often ask questions about overall meaning and purpose as well as questions about day-to-day mood. They ask whether participants have had trouble getting going that week, for instance, or whether they have felt like failures, or whether they feel hopeful about the future. And it seems perfectly reasonable to assume that someone *with* children under his or her roof, as opposed to someone whose children were taken away, would answer these questions with more optimism. The former have reasons to get up in the morning, reasons to feel like they have made something of their lives, reasons to be connected to the future.

During one of our conversations, Beth, the divorced schoolteacher from the previous chapter, told me that when her son, Carl, was at his most rejecting (refusing to see her, not responding to her texts), she made an effort to give him space and stopped communicating with him for a while. "And I was more miserable," she said, "than when I texted him and didn't hear back." It felt uncomfortable not to reach out to him. It felt uncomfortable not to give love.

In *Flow*, Csikszentmihalyi reports something similar. He finds that for single, non-churchgoing people, Sunday mornings are the low point of their week, for the simple reason that they have no demands placed on their attention. "For many, the lack of structure of those hours is devastating," he writes. Viktor Frankl, the psychiatrist and Holocaust survivor, talks about melancholy Sundays too, in his best-known book, *Man's Search for Meaning.* He calls it "Sunday neurosis" and defines it as "that kind of depression which afflicts people who become aware of the lack of content in their lives when the rush of the busy week is over." His therapeutic recommendation in situations of such distress, always, is to add meaningful activity to a life. That activity doesn't need to be pleasurable. It can even open up a person to pain. That isn't the point. The point is to have a reason to keep going. "If architects want to strengthen a decrepit arch," he notes, "they *increase* the load that is laid upon it, for thereby the parts are joined more firmly together." Therapists handling despairing patients, he therefore counsels, should "not be afraid to create a sound amount of tension through a reorientation toward the meaning of one's life."

And that's what choosing parenthood does: gives strength and structural integrity to one's life through meaningful tension.

If one takes meaning into consideration, happiness might best be described as "a zest for life in all its complexity," as Sissela Bok writes in her book. To achieve it means to "attach our lives to something larger than ourselves." To be happy, one must *do.* It could be something as simple as teaching Sunday school or as grand as leading nonviolent protests. It could be as cerebral as seeking the cure for cancer or as physical as climbing mountains. It could be creating art. And it could be raising a child—my "best piece of poetrie," as Ben Jonson said in his elegy for his seven-year-old son.

the remembering self

In an evening dads' class in St. Paul, a fellow named Paul Archambeau has the floor. He's different from the other fathers in the room. While most are first-time dads—or have, say, a three-year-old and a newborn—Paul has four children. His youngest is three, which makes him eligible for ECFE, but his oldest is eleven. "Now that Ben and Isaac are growing up," he says, "I long for the days when they would sit at the counter and eat Cheerios with their hands. Norah"—his youngest—"can drive me crazy with some of the things she does. But I just know, a year or two or three from now, I'm going to say, 'Man, that was fun.'"

Another father, Chris, whose son is seventeen months, looks surprised. "Why do you long for that?" he asks. "Because I see being able to play catch . . ."

Says another: "Yeah, I fight the feeling off every day—*I can't wait until he's older.*"

"I don't know," Paul admits. "Maybe the finality to it, knowing I'm never going to get those years back. Or maybe because I'm forgetting how hard it was."

The group debates this for a while.

"But here's the thing," says Paul. "I would bet that if someone did a study and asked, 'Okay, your kid's three, rank these aspects of your life in terms of enjoyment,' and then, five years later, asked, 'Tell me what your life was like when your kid was three,' you'd have *totally* different responses."

WITH THIS SIMPLE OBSERVATION, Paul has stumbled onto one of the biggest paradoxes in the research on human affect: we enshrine things in memory very differently from how we experience them in real time. The psychologist Daniel Kahneman has coined a couple of terms to make the distinction. He talks about the "experiencing self" versus the "remembering self."

The experiencing self is the self who moves through the world and should therefore, at least in theory, be more likely to control our daily life choices. But that's not how it works out. Rather, it is the remembering self who plays a far more influential role in our lives, particularly when we make decisions or plan for the future, and this fact is made doubly strange when one considers that the remembering self is far more prone to error: our memories are idiosyncratic, selective, and subject to a rangy host of biases. We tend to believe that how an episode ended was how it felt as a whole (so that, alas, the entire experience of a movie, a vacation, or even a twenty-year marriage can be deformed by a bad ending, forever recalled as an awful experience rather than an enjoyable one until it turned sour). We remember milestones and significant changes more vividly than banal things we do more frequently. And how long an activity lasts seems to have little influence on our recollections at all—two weeks of vacation, Kahneman noted in a 2010 TED lecture, won't be recalled with much more fondness or intensity than one week, because that extra week probably won't add much new material to the original memory. (Never mind that the experiencing self might really enjoy that extra week of vacation.)

In that same lecture, Kahneman confessed that the outsize power of the remembering self mystified him. "Why do we put so much weight on memory relative to the weight we put on experiences?" he asked the audience. "This is a bit hard to justify, I think."

But perhaps the answer is obvious: *children*. The remembering self ensures that we'll keep having them. More than almost anything else, the experience of parenthood exposes the gulf between our experiencing and remembering selves. Our experiencing selves tell researchers that we prefer doing the dishes—or napping, or shopping, or answering emails—to spending time with our kids. (I am very specifically referring here to Kahneman's study of 909 Texas women.) But our remembering selves tell researchers that no one—and nothing—provides us with so much joy as our children. It may not be the happiness we live day to day, but it's the happiness we *think* about, the happiness we summon and remember, the stuff that makes up our life-tales.

This is precisely what Paul tells his fellow dads. "Here's the best way to describe it," he says. "I was at the high school hockey tournament this weekend with all the kids. And it was kind of crazy. Especially with a three-year-old. Trying to sit and whatever. And at one point this woman going up the stairs was like, 'Are *all* those kids yours?'" He mimes her incredulity and pretends to point to four imaginary children. "And I'm like, 'Yeah.'" This he says ruefully, with an implied eye roll. "But then, I'm like"—and he pauses to reconsider—"*yeah.*" Marveling this time. "So it just seems to me," says Paul, "that when I'm in the moment, it can be chaotic. But if I can be shaken out of that moment, even if it's just for like a split second, it's like, you know what? This is a cool thing."

He just has to step away from the moment to see it.

Which isn't surprising. Lots of parents will tell you that when they aren't fighting with their teenagers about homework or scraping up raisins their toddlers have expertly ground into the kitchen floor, they're quite happy, upon reflection. In a 2007 poll taken by Pew, 85 percent of all parents rated their relationships with their minor children as most important to their personal happiness and fulfillment—more than relationships with their spouses, their parents, or their friends, and more than their jobs. When asked to *think* about what makes us happy, the answer is clear: our kids.

Csikszentmihalyi told me something similar when I interviewed him in Philadelphia. It's true that when he monitors people in real time, they're less apt to say they're in flow while with their children. But if you ask mothers to *recall* their greatest flow moments, the ones they're most likely to report are ones involving their kids. "Especially things like reading books to them," he said, "and seeing them pay attention, getting interested in things."

"In our interviews," says Dan P. McAdams, a psychology professor at Northwestern University, "there's a section where we focus on high points, low points, and turning points." McAdams studies how human beings form their identities through the stories they tell about themselves. He's spoken to hundreds of adult men and women, collecting their narratives, looking for patterns. "And the most common high point for midlife adults," he tells me, "is the birth of a first child." That's true for both men and women.

Storytelling, as Kahneman likes to say, is our natural response to memory. The episodes we recall become part of our identities, the delicate composition of who we are. Our remembering selves are in fact *who* we are, he goes so far as to say in *Thinking, Fast and Slow*, even though our experiencing selves do our actual living for us.

And if that's the case—if we *are* our remembering selves—then it matters far less how we feel moment to moment with our children. They play rich and crucial roles in our life stories, generating both outsize highs and outsize lows. Without such complexity, we don't feel like we've amounted to much. "You don't have a good story until something deviates from the expected," says McAdams. "And raising children leads to some pretty unexpected happenings."

Our stories may not always be pleasant as they're being lived. They can in fact be just the opposite, acquiring a warm hue only in retrospect. "I think this boils down to a philosophical question rather than a psychological one," Tom Gilovich, a professor of psychology at Cornell, tells me. "Should you value moment-to-moment happiness more than retrospective evaluations of your life?" He says he has no answer for this, but the example he offers suggests a bias. He recalls watching TV with his children at 3:00 A.M. when they were sick. "I wouldn't have said it was too fun at the time," he says. "But now I look back on it and say, 'Ah, remember the time we used to wake up and watch cartoons?'"

legacies

Kids do not just provide us with stories about ourselves. They give us a shot at redemption. McAdams, who's spent twenty-five years thinking about life stories, says that the most "generative" adults in his samples—the ones most concerned about bequeathing something meaningful to the next generation—are more inclined to tell stories about renewal and reinvention. Highly generative adults, he writes,

> invest considerable time, money, and energy into ventures whose long-term payoff is hardly a sure bet. Raising children, teaching Sunday School, agitating for social change, working to build up valued social institutions—these kinds of generative efforts often involve as much frustration and failure as fulfillment. Yet, if one's internalized and evolving life story—one's narrative identity—shows again and again that suffering can be overcome, that redemption typically follows life's setbacks and failures, then seeing one's life in redemptive terms would appear to be an especially adaptive psychological thing to do.

Children can often play a role in our life-redemption narratives. McAdams says he frequently hears from fathers, "If it hadn't been for my child, I'd still be unfocused and running around." And children often play the biggest role in the stories of those whose lives as parents are hardest: poor women. In *Promises I Can Keep,* an ethnography of young, single mothers by Kathryn Edin and Maria Kefalas, the authors write that "the redemptive stories our mothers tell speak to the primacy of the mothering role, how it can become virtually the only source of identity and meaning in a young woman's life." Absent better economic and marital prospects, the interviewees in Edin and Kefalas's book say that their children saved them, sparing them from more self-destructive lives.

Because they are lucky enough to have options—and therefore more than one way to create a meaningful life—the middle class may feel more constrained once their children come along, as if their lives have suddenly been condensed into a teacup. But children expand these parents' lives too. Kids open windows to new activities and new ideas and "bring different worlds to your home," as Philip Cowan puts it. They become obsessed with chess, for example, and you've never played it; they start learning about Islam at school, and you've never formally studied it—now the evening news makes a bit more sense. Their competence at things you can't do, and their mastery of subjects you know nothing about, can engender a wild sense of pride. Think of Nancy Darling staring at her son playing the violin, or Gayle, from the previous chapter, marveling at her daughter actually knowing who Erik Erikson was. "That's the kind of thing you live for," she'd said. "You want them to be better than you."

This pride need not come from kids' accomplishments either. It can stem—and often does—from their simple transformation into moral, compassionate creatures. All children start their lives as tiny narcissists. Yet somewhere along the line, almost without your noticing, they begin to appreciate suffering and want very much to assuage it. They bring you soup when you get sick. They tell you about keeping their mouths shut at lunch, because their friends were discussing a birthday party and not everyone present was invited. And you realize that all the love you showed along the way, all the lectures you gave about compassion and grace and respect—it has somehow all managed to stick.

McAdams has noticed a commonality among the stories from his most generative adults. They were consciously telling their tales to a younger generation, seeing them as parables their children could learn from. "It comes out as, 'I've developed a story for my life, replete with wisdom and folly, and this narrative exists as something I tell my kids about,'" he says. "My narrative identity can have an effect on other people."

The most productive, generative adults see their children as their superegos, in other words. Their kids hover over them and guide all of their moral choices. If these adults falter or behave ignobly, they know their kids will see; the same is true if they do well. They are exquisitely aware of themselves as role models. They know they are being watched.

This isn't, in McAdams's experience, how everyone thinks. Roughly one hundred years ago, Freud observed that many people spend their time reenacting the dramas of their pasts, seeking the approval of ghosts. They think of their *parents* as their superegos, the imaginary judges they've constantly got to please. But this is not true of the adults who are most concerned about leaving a lasting legacy. In their eyes, "the evaluator shouldn't be the past generation," says McAdams. "It should be the next." They are freed up to invent their own lives, knowing they won't be governed by the norms of a previous generation. They want their children to be their final judges.

"NONCALORIE CHOCOLATE" IS WHAT the social psychologist Daniel Gilbert calls his granddaughters. "They're all the joy and as much fun as you want," he says, "and none of the responsibility."

But Sharon's circumstances are different. Her grandson *is* her child—both functionally and now in the eyes of the law. She adopted him after Michelle died.

Margaret Mead talked about the helplessness of modern American parents, who, without years of folkways to guide them, were at sixes and sevens about how to raise their children—vulnerable to fads, untrusting of their own instincts, suspicious that anything their own mothers knew about child-rearing was bound to be out-of-date.

One can only begin to imagine how acutely Sharon must have felt these anxieties throughout her parenting life. She had little information to go on about how to raise a child with depression, and little to go on about how to handle an adopted child with cognitive and behavioral problems. Her whole young motherhood was a master class in winging it. And then, decades later, she found herself raising a child all over again—Cam—only to discover that many of the rules and habits she'd learned the first time around were again useless. Now it wasn't okay to leave a child unattended in a car for five seconds to go grab something from the store. Now opening a stroller required two hands and a foot. Now all the experts were telling her to get down on her hands and knees—as a senior citizen!—and engage in intensive play, rather than simply encouraging her to tell her grandson to go entertain himself.

But Sharon's circumstances have *always* been different. Her life hasn't just involved extemporizing her way through parenting. It has also involved extemporizing her way through grief, and far too much of it, mourning not just one child but two. Losing a loved one, like having a child, is another one of life's abrupt transitions for which one can never adequately prepare. And now Sharon finds herself extemporizing through yet another abrupt transition.

I wouldn't have known if I hadn't phoned, nearly two years after meeting her, to let her know that my book was almost done. When I reached her—it took several attempts, which should have been a clue—she sounded very tired, but underneath the fatigue was the same Annie Oakley toughness. "Well," she said, "my situation has changed a bit since we've seen one another. . . ." And then she shared her news.

Sharon is dying. It's cancer, it's in her brain, it's quick and aggressive. She talked calmly about it. "You can't be a person of faith," she said, "and not have thought a lot about dying." For a few months, she experienced no pain, and she tolerated the chemo very well. She was so well connected—through her church, through her friends at ECFE, through years of living in the same neighborhood—that she and Cam never wanted for company or help or home-cooked meals.

But then her short-term memory started to fade, and there were complications from her treatment; it became all too clear that she was no longer fit enough to handle a small boy. So she reorganized her life. She made plans to move to the same city as her one remaining grown child, to whom she's still close. And she arranged for a younger relation to take Cam in, and it's with this relation that Cam will probably stay once Sharon is gone. He'll be part of a family with children who are still at home; they all love Cam and Cam loves them.

Parents do not, if they are lucky, have to grapple with the fact of their own mortality from day to day. But if they are forced to, as Sharon has been, something can happen. The clarity of their role, rather than its complexity, locks into view. Meeting daily obligations, arranging for the future, communicating unconditional and eternal love—these become the primary tasks of a parent who is dying. They are the primary tasks of a healthy parent too, but the snowy static of the outside world often makes them difficult to see. In an award-winning essay, the writer Marjorie Williams, who was also diagnosed with cancer while her children were young, talked about this: "Having found myself faced with that old bull-session question (What would you do if you found out you had a year to live?), I learned that a woman with children has the privilege or duty of bypassing the existential. What you do, if you have little kids, is lead as normal a life as possible, only with more pancakes."

I spoke to Sharon the week that Cam was set to leave. I caught her at home—she was now always at home—and she and Cam were sitting in the living room together, watching *Curious George*. When Cam briefly walked out of earshot, she told me how he was doing. "He's been showing a lot of anger," she said. "There was a day when he picked up a shoe and threw it at my head. He knew that that's where the cancer is." But he also knew that Sharon didn't choose to have cancer. Even at four and three-quarters, Cam could make that distinction. What his anger provided was a glimpse into how much he'd miss her, and a chance for Sharon to explain that love doesn't stop once a person dies, and neither does parenthood. "He's also very loving, saying, 'I love you forever and ever and ever,'" said Sharon. "We talk to each other a lot about loving each other forever and ever and ever. And that even without seeing each other, I'll be his mommy and he'll be my son."

I wondered if she felt guilt. "I do," she said. "I feel like I'm abandoning Cam." But then she said something I will never forget. She said she also felt relief. "Now he's going to have *two* adults who love him and will be taking care of him for the rest of his growing up. It's very comforting. It's a better life for him than he would have had if he'd stayed with me." Sharon didn't think she'd have had the courage to make that choice for Cam if she'd remained well.

For the moment, she said, she was just trying to savor their last few days together in her beautiful old house. "I'm trying to be present," said Sharon. "That's all I can do. That," she added, "and spending a lot of time watching *Curious George.*"

Which is just as Marjorie Williams had said. Normal life, but with more pancakes.

Kids may complicate our lives. But they also make them simpler. Children's needs are so overwhelming, and their dependence on us so absolute, that it's impossible to misread our moral obligation to them. *It's for life,* as Sharon says. But it also *is* our lives. There's something deeply satisfying about that. Williams wrote that motherhood gave her permission to circumvent existential questions when she fell ill, and perhaps that's true. But I suspect that parenthood helped reduce the number of existential questions she had in the first place. She knew what she had to do each day, and why she was here. And the same was true of Sharon. Even at her weakest—even when she was well past the point of charging through the sprinklers at the splash pad or hoisting Cam onto a jungle gym—she knew exactly what she was supposed to be doing with her last moments of strength. She was supposed to be watching *Curious George* with Cam.

And when she dies, a member of her family will do for Cam exactly what Sharon did for Cam's mother all those years ago: sign on for life. It seems to be a recurring theme in Sharon's family, a sacred code of conduct that they all share in the happiest and worst of times. This is what parents do—what all of us do, in fact, when we're at our unrivaled best. We bind ourselves to those who need us most, and through caring for them, grow to love them, grow to delight in them, grow to marvel at who they are. Gift-love at its purest. Even in the midst of pain and loss, it is, miraculously, still possible to summon.

acknowledgments

Writing your first book is not unlike the early days of raising your first child. You're awed by the magnitude and meaning of this new undertaking, certainly, but also housebound, perpetually preoccupied, and (perhaps worst of all) presumed to be competent at something you know essentially nothing about. It takes a huge network of friends and family and colleagues to make such a project work.

There is, for starters, Tina Bennett, who isn't just a brilliant champion of writers' ideas, but (secretly) a brilliant editor of them; she also has a genius for friendship, which I enjoyed long before enjoying her genius as an agent. Her colleague Svetlana Katz is a model of effortless professionalism.

At Ecco, Lee Boudreaux swept up this project with so much enthusiasm that her energy alone could have powered my laptop. As an editor, she's part of a dying breed: someone who pays close attention not just to individual sentences but big ideas; someone who talks through chapters as well as rereads them, endlessly. That she also happens to be one of the funniest and loveliest women to hang out with seems almost too good to be true, but there you have it. I'm also grateful to her for making me look like I talk at a normal speed.

The whole Ecco team is tremendous: Dan Halpern, the publisher, gave me the freedom and (toward the home stretch) extra time to get it right; Michael McKenzie, Ecco's publicity director, knows the media almost as well as us journalists; and Ashley Garland gave good PR guidance at every step of the way. Art director Allison Saltzman produced the perfect, playful cover. And thanks, too, to Ryan Willard, Andrea Molitor, Craig Young, and Ben Tomek for everything they've quietly done to keep this strange process running so smoothly.

I could not have written this book without the full support of Adam Moss and Ann Clarke at *New York* magazine. Maybe there are other employers out there who allow their employees two years off from their jobs, but if there are, I haven't heard about them; for a while, Adam and Ann had me more or less convinced I was living in Sweden. Adam also published the magazine story that formed the original basis for this book. A couple of paragraphs from it—as well as a couple of paragraphs from a subsequent story I did about the enduring effects of high school—appear here.

At *New York,* it was Lauren Kern (now at the *Times*) who heard me pitch the original magazine story, thought it was swell, and edited it into a readable state. Throughout my career, I've been lucky to work with editors who've made my writing much better, including John Homans, Vera Titunik, Al Eisele, Marty Tolchin, David Haskell, Ariel Kaminer, and Mark Horowitz (more on that last guy in a bit). David and Ariel were invaluable early readers, too, providing terrific comments and suggestions (Ariel read not just early, but often); so were the amazing Bob Roe, Kyla Dunn, and Caroline Miller (who's done wonders for my understanding of adolescents, and first hired me at *New York* magazine in 1997). My friend Josh Shenk talked me through the early stages of this book. My colleague Chris Smith provided crucial advice at the end. My colleague Bob Kolker troubleshot this book over so many lunches that he needn't have bothered reading the final product, he knew it so well. Elaine Stuart-Shah helped a great deal with the initial research; Rachel Arons had a singular talent for archival spelunking; Rob Liguori made fact-checking this book look easy, which Lord knows it wasn't. I thank him for the many times he artfully rescued me from myself.

Though I mainly quote published research in this book, many scholars also made time to speak to me by phone, in person, or by long email volley, including David Dinges, Michael H. Bonnet, Mimi Ito, Linda Stone, Mary Czerwinski, Roy F. Baumeister, Matthew Killingsworth, Arthur Stone, Dan P. McAdams, Mihaly Csikszentmihalyi, David E. Meyer, Tom Bradbury, Susan McHale, Mike Doss, Kathryn Edin, Alison Gopnik, Sandra Hofferth, Andrew Cherlin, Steven Mintz, Dalton Conley, Kathleen Gerson, E. Mark Cummings, Clay Shirky, Brené Brown, Gerald R. Patterson, Donald Meichenbaum, Arnstein Aassve, Ann Hulbert, and Andrew Christensen. I owe all of them my heartfelt thanks. I feel especially indebted to Dan Gilbert, George Vaillant, Robin Simon, Nancy Darling, Larry Steinberg, B. J. Casey, and Carolyn and Philip Cowan, all of whom went out of their way on my behalf.

When I was struggling to find a methodical, dignified approach to pulling together a sample of parents, it was Bill Doherty at the University of Minnesota who suggested I try ECFE (a remarkable state education program). He and his daughter Elizabeth hooked me up with Annette Gagliardi and Todd Kolod, who shared their wisdom, planned my trip, and graciously allowed me to attend many of their classes; Barb Dopp, Kathleen Strong, Valerie Matthews, and Kristine Norton let me sit in on theirs as well. It was my pal Shaila Dewan who wisely advised me to go to Sugar Land and Missouri City, on account of their shifting demographics; Mimi Swartz sent me to Kathryn Turcott and Rallou Matzakos, both of whom plugged me into the PTO at Palmer Elementary in Missouri City. Mimi and Lisa Gray and Amy Weiss helped orient me in Houston, for which I can't thank them enough.

Nor, of course, can I ever sufficiently thank all the families who gamely participated in this book: Angie and Clint Holder; Jessie and Luke Thompson; Marta Shore; Chrissy Snider; Paul Archambeau; Laura Anne Day; Leslie Schulze; Steve and Monique Brown; Lan Zhang; Cindy Ivanhoe; Carol Reed; Angelique Bartholomew; the moms and dads who shared their thoughts during ECFE classes; the moms and dads of teens who shared their stories in full. You were generous, you were warm, you took a flier on a total stranger. You spoke honestly about intense subjects and intensely about what that honesty meant. This was especially true of Sharon Bartlett, as inspiring a woman as I've ever known. She died on July 9, 2013. Her surviving daughter has clearly inherited the generosity of her mother. I suspect Cam will too.

If Sharon taught me anything, it was the necessity of friendship and community. A number of people in my life made the loneliness of book-writing bearable, not just by boosting my morale but by spitballing ideas so that I might feel less alone with them. So thanks to: Sarah Murray; Nina Teicholz and Gregory Maniatis; Mikaela Beardsley; Sue Dominus and Alan Burdick; Steve Warren; Brian Baird; Rebecca Carroll; Brian Hecht and Doug Gaasterland; Fred Smoler and Karen Hornick; Josh Feigenbaum; Doug Dorst; Thom Powers and Raphaela Neihausen; Howard Altmann; Dimple Bhatt; Julie Just and Tom Reiss; and Eric Himmel.

One cannot write a book about parenthood without rethinking one's own parents. Norman and Rona Senior had me when they were young, *so* young, making trade-offs and compromises I still cannot imagine although I'm a parent myself. Their love and unconditional support are what launched me into the world and still power me through it. In Ken Senior and Deanna Siegel Senior (another early and wonderful reader), I'm pretty sure I have the most loving and least complicated relationship a human being can have with a brother and sister-in-law; I don't think I'd have written this book without their friendship, feedback, and babysitting. Jon Sarnoff and Allison Soffer may as well be my brother and sister. Thanks to them—as well as their spouses, Ellen Lee and Bob Soffer—I've learned what mothering and fathering looks like at its best; how I wish their mom were here to see her children parent as she did. (And thanks to Dylan, Max, Miles, Mia, Ben, and Caroline, Rusty will never want for cool cousins, who'll be just as close, with any luck, as all of us.) Sam Budney and Stella Samuel aren't related by blood, but they may as well be, and they made my life possible, as well as my son's. George and Eleanor Horowitz aren't related by blood either, but I admire them, would take a bullet for them, feel a stubborn connection to them that takes me by surprise; I feel so grateful for the many efforts they've made, because blended families are, no matter how much you work at them, really hard.

And then there's Mark Horowitz, who stole my heart one day when he declared that sometimes we did things in life—made sacrifices, took risks—simply because we loved someone, and that was reason enough. He taught me how to write; he taught me how to be a wife; he showed me the importance of old-fashioned concepts like duty and honor. He cooked several hundred meals for me as I was writing this book, and cooked up several hundred ways to improve my writing too. Together, we had Rusty, to whom this book is dedicated—as is, in large measure now, my life. Without this kid, the world wouldn't be half so beautiful, or half so meaningful, or half so large. How I love you, darling boy. You'll never know the half of it, and that's just fine.

a conversation with jennifer senior and curtis sittenfield

CS: It's difficult to describe *All Joy* without making it sound like a how-to parenting manual. How do you get around this challenge?

JS: Thank you for noticing! I generally tell people that it's a book about how children affect their parents, not the other way around, which means I'm sort of examining the habits of parents as a species, as if they were frogs or bats.

CS: As a journalist, you've written about a wide range of topics, including pop culture and politics, so I'm wondering why parenthood is the subject that elicited a book from you.

JS: You're right, and were this a parenting book, it wouldn't even occupy the same hemisphere as the other pieces I've done. (Here's where I make an embarrassing confession: I have purchased exactly one parenting book in my lifetime.) But I consider this a social science book, and I've done plenty of social science stories over the years: about the psychological effects of high school on our adult years; about loneliness and cities; about burnout; about our culture-wide obsession with happiness. Also, I sort of think of this book as a series of mini-ethnographies—little portraits of how American families live now—and that comes pretty naturally. I was an anthro major, and even when I wrote about the Senate, which used to be often, I always treated it as an other-planetary universe with its own folkways.

CS: This book has its origins in a much-buzzed-about *New York* magazine cover story. In that article but not in the book, you discussed your own experiences as a parent. Why didn't you include yourself in the book? Can you share a bit about your family?

JS: You know what's weird about that? I mentioned myself and my son in just two paragraphs of that story, but because they were the *first* two paragraphs, people forget, thinking I'd sprinkled my own experiences throughout. The only reason I brought myself up—both early in the magazine story and in this book—was to alert readers that I was writing as a parent myself; I wanted them to know who their narrator was. But the specifics of my own story seem irrelevant, or unimportant, or at the very least too idiosyncratic from which to generalize. You get a fuller picture when you (a) look at the full spectrum of social science research about families, and (b) talk to a broad range of moms and dads.

For the record, though: My husband and I have one son, who's seven, and my husband has two grown kids from a previous marriage, though I entered their lives when they were adolescents.

CS: One of the book's fascinating tidbits is the implication that parents have friction with teens in some sense because the parents are jealous—is that an accurate interpretation?

JS: The effects of adolescent children on their parents are wildly under-discussed, but I think jealousy is only a small part of it. (Though I am amazed by the work of Laurence Steinberg, the psychologist who found that fathers get especially depressed when their teenage sons start to date.) What generally seems to happen is that adolescent kids make their parents take stock of every life choice they've ever made—their marriages and careers especially. Adolescents can be so rejecting and critical that they expose all the holes in their parents' lives: *Well, now that my kid's dispensed with me, all I have is my marriage and my job, and I'm not thrilled with either.* And because teenagers are proto-adults, they occasion all this second-guessing about parenting choices: *Wow, my kid has turned into quite a yeller. Did I turn her into that? Is she copying me?*

CS: You suggest that many moms would be better off being more like dads. Can you explain what you mean?

JS: I mean this only in the sense that fathers, at the moment, seem less frantically perfectionist about their parenting than mothers do, probably because they aren't burdened by the same unattainable cultural ideals (whether they're real, like the Tiger Mom, or fictional, like June Cleaver). It's a crude generalization, yes, and of course there are exceptions. But I was struck—both in conversations and in the data—by how much less pressure fathers felt to be playing with their children every free moment they had, and how much quicker they were to claim their right to free time. And I'm talking about *good* fathers, *engaged* fathers. If mothers did the same, one wonders what would happen—*Hey, glad you're back from that bike ride, now I'm going to the gym!* It's just possible that domestic divisions of labor might shift a little in their favor. And considering moms spend more time with their kids than they ever have (including the 1960s, when most women were still at home), it's just possible this generation of children will be no worse off for it.

reader's guide

1. Senior suggests that there are three developments that have changed and complicated modern parenting in America: first, control over when and how many children to have; second, how our work lives have become more complicated; and third, the transformation of the child's role in home and society. Do you find these three issues present in your own household? How have you found ways to address these challenges in parenting?

2. In her chapter on marriage, Senior says men and women experience time differently. Women, she notes, feel like their time is divided and subdivided, and their failure to get everything done in the amount of time they have is tied to their sense of self. Senior writes, "Being compelled to divide and subdivide your time doesn't just compromise your productivity. . . . It also creates a feeling of urgency—a sense that no matter how tranquil the moment . . . there's always a pot somewhere that's about to boil over" (page 59). Do you find that this aligns with your role as a parent? Does this affect your relationship with your partner in the way Senior suggests?

3. Senior says that one of the most common fights between parents is the household division of labor. Is this true in your house? Did

you arrange these divisions ahead of time or improvise as you went along?

4. In her chapter "Adolescence," Senior suggests that having adolescent children around unmasks tensions that were previously concealed. Did you find this to be the case? If so, what tensions? Did it start earlier than adolescence, even?

5. Senior suggests that one of the joys of having young children is re-engaging with the physical world—doing things with your hands. Did this happen for you? What did you do? What pleasure does (or did) your young child allow you to have that you wouldn't have done otherwise?

6. Senior observes that kids give adults a chance to practice philosophy again. What are some of the most baffling questions your children have asked you? How did you answer them? Did you find it a pleasure to answer those questions?

7. Do you think that the value of happiness is overemphasized in our culture, and we all might relax a little if we embraced the notions of duty and service instead?

8. What parts of parenting do you find hardest to talk about? What parts give you most pleasure?

9. As Senior profiles Angie's and Clint's daily challenges as modern parents, we see Clint declare: "I am the standard." While Senior observes that fathers have fewer cultural and historical scripts to follow, they do tend to judge themselves less harshly than mothers, which suggests mothers might have something to learn from them. Do you agree with this suggestion? How would you go about it?

10. In what ways does Senior's description of joy mirror your own experience as a parent, and in what ways does it differ? Does Senior's distinction between joy and fun capture your own reality as a parent? How so?

11. Sharon and Cam's story becomes one of the book's most poignant threads. How did Sharon's unique perspective as both a parent and grandparent affect you?

12. In her chapter "Concerted Cultivation," Senior speaks of the "over-scheduled parent," asking, "What, precisely, are we preparing our children for? How, as mothers and fathers, are we supposed to prepare them for it? Have parents always operated this blindly? Or were the roles of parents and children more clearly—and simply—defined in the past?" (page 123). Do you believe the "globalized future" has affected your choices as a parent? How so? How do you deal with these stressors in your family?

13. As a journalist, Senior pairs a great deal of research and complex social science with intimate stories of real families. Did you find this approach illuminating? Insightful? Relatable?

14. How do you think your "remembering self" will look back on the years of raising a toddler? A middle-schooler? An adolescent?

notes

introduction

4 **"fragile and mysterious"** Alice S. Rossi, "Transition to Parenthood," *Journal of Marriage and Family* 30, no. 1 (1968): 35.

4 **What was the effect of parenthood on *adults?*** Ibid., 26.

4 **"We knew where babies came from"** E. E. LeMasters, "Parenthood as Crisis," *Marriage and Family Living* 19, no. 4 (1957): 352–55, 353.

4 **"Loss of sleep"** Ibid., 353–54.

5 **economic pressure, less sex, and "general disenchantment"** Ibid., 354.

5 **another landmark paper** Norval D. Glenn, "Psychological Well-being in the Postparental Stage: Some Evidence from National Surveys," *Journal of Marriage and Family* 37, no. 1 (1975): 105–10.

5 **children tended to negate its effects** Paul D. Cleary and David Mechanic, "Sex Differences in Psychological Distress Among Married People," *Journal of Health and Social Behavior* 24 (1983): 111–21; Sara McLanahan and Julia Adams, "Parenthood and Psychological Well-being," *Annual Review of Sociology* 13 (1983): 237–57.

5 **Throughout the next two decades** Recent papers showing this phenomenon include: David G. Blanchflower and Andrew J. Oswald, "International Happiness: A New View on the Measure of Performance," *Academy of Management Perspectives* 25, no. 1 (2011): 6–22; Robin W. Simon, "The Joys of Parenthood Reconsidered," *Contexts* 7, no. 2 (2008): 40–45; Kei M. Nomaguchi and Melissa A. Milkie, "Costs and Rewards of Children: The Effects of Becoming a Parent on Adults' Lives," *Journal of Marriage and Family* 65 (May 2003): 356–74.

5 **ranked sixteenth out of nineteen** Daniel Kahneman et al., "Toward National Well-being Accounts," *American Economic Review* 94, no. 2 (2004): 432.

5 **In an ongoing study** Killingsworth uses an iPhone app to track people's emotions as they go about their daily lives. For more details about his project, see http://www.trackyourhappiness.org; to see published material derived from this data set, see Matthew A. Killingsworth and Daniel T. Gilbert, "A Wandering Mind Is an Unhappy Mind," *Science* 330, no. 6006 (November 2010): 932.

5 **"Interacting with your friends is better"** Daniel A. Killingsworth, interview with the author, February 6, 2013.

6 **more *highs* as well as more lows** Angus Deaton and Arthur Stone, "Evaluative and Hedonic Wellbeing Among Those With and Without Children at Home," *Proceedings of the National Academy of Sciences* 111, no. 4 (2014): 1328–1333.

6 **greater feelings of meaning and reward** The most comprehensive and forward-thinking of these studies is by Debra Umberson and Walter Gove, "Parenthood and Psychological Well-being: Theory, Measurement, and Stage in the Family Life Course," *Journal of Family* Issues 10, no. 4 (1989): 440–62.

6 **"high-cost/high-reward activity"** William Doherty, interview with the author, January 26, 2011.

6 **as much as being legally drunk** Michael H. Bonnet, interview with the author, November 17, 2011.

7 **adults often view children as one of life's crowning achievements** See, for example, Andrew J. Cherlin, *The Marriage Go-Round: The State of Marriage and the Family in America Today* (New York: Vintage Books, 2010), 139.

7 **between the ages of twenty-five and twenty-nine** US Department of Commerce and Office of Management and Budget, *Women in America: Indicators of Social and Economic Well-being* (March 2011), 10.

7 **resulted from assisted reproductive technology** Centers for Disease Control and Prevention, "Assisted Reproductive Technology," available at: http://www.cdc.gov/art/ (accessed August 15, 2014). Today, assisted reproductive technology is responsible for over 1 percent of American babies, a figure that's more than doubled in the last decade.

8 **"some at inauspicious times"** Jerome Kagan, "Our Babies, Our Selves," *The New Republic* (September 5, 1994): 42.

8 **34 percent of women** Bureau of Labor Statistics, *Women in the Labor Force: A Databook,* report 1034 (December 2011), 18–19.

9 **male comic, Louis C.K.** "Louis C.K. on Father's Day," June 20, 2010, available at: http://www.cbsnews.com/video/watch/?id=6600481n (accessed April 4, 2013).

10 **"economically worthless but emotionally priceless"** Viviana Zelizer, *Pricing the Priceless Child* (New York: Basic Books, 1985), 14.

10 **every three minutes** Cheryl Minton, Jerome Kagan, and Janet A. Levine, "Maternal Control and Obedience in the Two-Year-Old," *Child Development* 42, no. 6 (1971): 1880, 1885.

11 **when teenagers fight most intensely with their parents** Brett Laursen, Katherine C. Coy, and W. Andrew Collins, "Reconsidering Changes in Parent-Child Conflict Across Adolescence: A Meta-analysis," *Child Development* 69, no. 3 (1998): 817–32.

11 **the most work-life conflict** Kerstin Aumann, Ellen Galinsky, and Kenneth Matos, "The New Male Mystique," in Families and Work Institute (FWI), *National Study of the Changing Workforce* (New York: FWI, 2008), 2.

12 **perhaps most recently** Judith Warner, *Perfect Madness: Motherhood in the Age of Anxiety* (New York: Riverhead Books, 2005), 20.

chapter one

15 **"I held the baby up to the light"** Melvin Konner, *The Tangled Wing: Biological Constraints on the Human Spirit* (New York: Henry Holt, 2002), 297.

16 **Minnesota's Early Childhood Family Education program** Mary Owen, Minnesota Department of Education, communication with the author, April 9, 2013.

17 **"I have not been alone in the bathroom since October"** Erma Bombeck, *Motherhood: The Second Oldest Profession* (New York: McGraw-Hill, 1983), 16.

18 **"human happiness is realizable on Earth"** John M. Roberts, "Don't Knock This Century. It Is Ending Well," *The Independent*, November 20, 1999.

18 **"Our lives become an elegy"** Adam Phillips, *Missing Out: In Praise of the Unlived Life* (New York: Farrar, Straus and Giroux, 2013), xiii.

19 **30.3 years old** Skip Burzumato, Assistant Director, National Marriage Project, communication with the author, March 27, 2013. See also Kay Hymowitz et al., "Knot Yet: The Benefits and Costs of Delayed Marriage in America," The National Marriage Project at the University of Virginia, 2013, 8, available at: http://nationalmarriageproject.org/wp-content/uploads/2013/03/KnotYet-FinalForWeb.pdf.

19 **"have their first child more than two years after marrying"** Hymowitz et al., "Knot Yet."

21 **the population seems to divide roughly in thirds** David Dinges, interview with the author, November 18, 2011.

21 **who'd had six hours of sleep or less** Daniel Kahneman et al., "A Survey Method for Characterizing Daily Life Experience: The Day Reconstruction Method," *Science* 306, no. 5702 (2004): 1778.

22 **"worth a $60,000 raise"** Ibid., 1779; Norbert Schwarz, Charles Horton Cooley Collegiate Professor of Psychology, University of Michigan, communication with the author, September 15, 2011.

22 **parents of children two months old and younger** National Sleep Foundation, "2004 Sleep in America Poll," March 1, 2004, available at: http://www.sleepfoundation.org/sites/default/files/FINAL%20SOF%202004.pdf (accessed May 6, 2013).

22 **parents of newborns average the same amount of sleep** Hawley E. Montgomery-Downs et al., "Normative Longitudinal Maternal Sleep: The First Four Postpartum Months," *American Journal of Obstetrics and Gynecology* 203, no. 5 (2010): 465e.1–7.

22 **"the effects of sleeping for four hours every night"** Michael H. Bonnet, interview with the author, November 17, 2011.

23 **"the more willpower people expended"** Roy F. Baumeister and John Tierney, *Willpower: Rediscovering the Greatest Human Strength* (New York: Penguin Books, 2012), 3, 33.

23 **fighting off the urge to sleep** Ibid.

24 **"Babies may be sweet"** Adam Phillips, *Going Sane: Maps of Happiness* (New York: HarperCollins, 2005), 66.

25 **"The modern child"** Ibid., 78.

25 **"Children would be very surprised"** Ibid., 79.

25 **"One of the most difficult things"** Adam Phillips, *On Balance* (New York: Farrar, Straus and Giroux, 2010), 33.

26 **distinction between a lantern and a spotlight** Alison Gopnik, *The Philosophical Baby: What Children's Minds Tell Us About Truth, Love, and the Meaning of Life* (New York: Farrar, Straus and Giroux, 2009), 129.

26 **"Anyone who tries to persuade a three-year-old** Ibid., 13.

27 **"Everybody would like to be in the present"** Daniel Gilbert, interview with the author, March 22, 2011.

30 **the work of the Hungarian psychologist** Mihaly Csikszentmihalyi, *Flow: The Psychology of Optimal Experience*, 1st paperback ed. (New York: HarperPerennial, 1991). All subsequent citations refer to this edition.

31 **"goal-oriented and bounded by rules"** Ibid., 49.

31 **"rules that require the learning of skills"** Ibid., 72.

31 **"This expansive lantern consciousness"** Gopnik, *The Philosophical Baby,* 129.

31 **"at the boundary between boredom and anxiety"** Csikszentmihalyi, *Flow,* 52.

32 **"To the extent that we are not maximally happy"** Daniel Gilbert, interview with the author, March 22, 2011.

32 **"setting aside a chunk of time"** Benjamin Spock, *Dr. Spock Talks with Moth-*

ers (Boston: Houghton Mifflin, 1961), 121, quoted in Ann Hulbert, *Raising America: Experts, Parents, and a Century of Advice About Children* (New York: Vintage Books, 2004), 353.

32 **"the most negative emotion I experienced"** Daniel Gilbert, interview with the author, March 22, 2011.

33 ***apart* from everyday life** Csikszentmihalyi, *Flow*, 58, 60, 158–59.

33 **"Let me tell you a couple of things"** Mihaly Csikszentmihalyi, interview with the author, July 25, 2011.

35 **roughly one-quarter of employed men and women** Bureau of Labor Statistics, "Work at Home and in the Workplace, 2010," TED: The Editor's Desk (blog), June 24, 2011, available at: http://www.bls.gov/opub/ted/2011/ted_20110624.htm.

36 **rats seeking pellets** For a modern comparison of the Internet and Skinner boxes, see Sam Anderson, "In Defense of Distraction," *New York Magazine* (May 17, 2009), available online at http://nymag.com/news/features/56793/; also Tom Stafford, "Why email is addictive (and what to do about it)," Mind-Hacks (blog), available online at http://mindhacks.com/2006/09/19/why-email-is-addictive-and-what-to-do-about-it/. For a thorough explanation of Skinner boxes, see, generally, B. F. Skinner, "The Experimental Analysis of Behavior," *American Scientist* 45, no. 4 (1957): 343–71.

37 **"Email apnea"** Linda Stone, communication with the author, April 11, 2013.

37 **"work from home *all the time*"** Dalton Conley, *Elsewhere, U.S.A.* (New York: Pantheon Books, 2008), 13, 29.

37 **we don't process information as thoroughly** Mary Czerwinski, interview with the author, June 8, 2011.

38 **"There's a warm-up period"** David E. Meyer, interview with the author, June 10, 2011.

38 **"This is over and above the stuff"** Ibid.

41 **you believe that women should stop getting in their own way** Sheryl Sandberg, *Lean In: Women, Work, and the Will to Lead* (New York: Knopf, 2013).

41 **much-discussed story about work-life balance** Anne-Marie Slaughter, "Why Women Still Can't Have It All," *The Atlantic* (July–August 2012).

42 **too many children running around** Andrew J. Cherlin, *The Marriage-Go-Round: The State of Marriage and the Family in America Today* (New York: Vintage Books, 2010), 44.

42 **Women gained a bit more control over their lives** Stephanie Coontz, *The Way We Never Were: American Families and the Nostalgia Trap* (New York: Basic Books, 1992).

42 **median age of women at first marriage fell to twenty** US Census Bureau, "Figures," at "American Community Survey Data on Marriage and Divorce," available at: http://www.census.gov/hhes/socdemo/marriage/data/acs (accessed April 22, 2013).

42 **birth rates increased** Coontz, *The Way We Never Were,* 24.

42 **women started dropping out of college** Betty Friedan, *The Feminine Mystique* (New York: W. W. Norton, 2001), 243.

42 **had evened out again** Ibid.

43 **"Regardless of their educational level"** Cherlin, *The Marriage-Go-Round,* 188.

43 **"get un-married and be free"** Claire Dederer, *Poser: My Life in Twenty-three Yoga Poses* (New York: Farrar, Straus and Giroux, 2011), 283.

43 **Americans have come to define liberty** Coontz, *The Way We Never Were,* 51.

44 **"learn to live somewhere"** Phillips, *Missing Out,* xi.

chapter two

45 **"My wife's anger toward me"** Barack H. Obama, *The Audacity of Hope* (New York: Vintage reprint edition, 2008), 531.

47 **83 percent of all new mothers** LeMasters, "Parenthood as Crisis," 353.

47 **90 percent of them experienced** Brian D. Doss et al., "The Effect of the Transition to Parenthood on Relationship Quality: An Eight-Year Prospective Study," *Journal of Personality and Social Psychology* 96, no. 3 (2009): 601–19.

47 **correlation between children and marital satisfaction** J. M. Twenge, W. K. Campbell, and C. A. Foster, "Parenthood and Marital Satisfaction: A Meta-analytic Review," *Journal of Marriage and Family* 65, no. 3 (2003): 574–83.

48 **indicated that their marriage was "in some distress"** Carolyn Cowan and Philip A. Cowan, *When Partners Become Parents: The Big Life Change for Couples* (New York: Basic Books, 1992), 109.

48 **more likely to be happy raising children as part of a couple** W. Bradford Wilcox, ed., "The State of Our Unions: Marriage in America 2011," National Marriage Project at the University of Virginia and the Center for Marriage and Families at the Institute for American Values, available at: http://www.stateofourunions.org/2011/index.php (accessed April 19, 2013).

48 **marital satisfaction curve bends noticeably** See, for example, Thomas N. Bradbury, Frank D. Fincham, and Steven R. H. Beach, "Research on the Nature and Determinants of Marital Satisfaction: A Decade in Review," *Journal of Marriage and Family* 62 (November 2000): 964–80; Daniel Gilbert, *Stumbling on Happiness* (New York: Vintage Books, 2007), 243 (chart).

48 **"By the time our children were in elementary school"** Cowan and Cowan, *When Partners Become Parents,* 2.

49 **92 percent of their sample couples** Ibid., 107.

49 **lesbian couples also showed increases in conflict** Abbie E. Goldberg and Aline Sayer, "Lesbian Couples' Relationship Quality Across the Transition to Parenthood," *Journal of Marriage and Family* 68, no. 1 (2006): 87–100.

49 **In 2009 an elegantly designed study** Lauren M. Papp, E. Mark Cummings, and Marcie C. Goeke-Morey, "For Richer, for Poorer: Money as a Topic of Conflict in the Home," *Family Relations* 58 (2009): 91–103.

49 **in another study, the same researchers** Lauren M. Papp, E. Mark Cummings, and Marcie C. Goeke-Morey, "Marital Conflicts in the Home When Children Are Present," *Developmental Psychology* 38, no. 5 (2002): 774–83.

49 **"When parents are *really* angry"** E. Mark Cummings, interview with the author, January 21, 2011.

53 **worked a full month extra** Arlie Russell Hochschild, *The Second Shift: Working Parents and the Revolution at Home* (New York: Penguin, 2003), 4.

53 **women are doing far less housework** Suzanne M. Bianchi, "Family Change and Time Allocation in American Families," *Annals of the American Academy of Political and Social Science* 638, no. 1 (2011): 21–44.

53 **updated introduction to her book** Hochschild, *The Second Shift,* xxvi.

53 **men's economic fortunes have fallen** Hanna Rosin, *The End of Men: And the Rise of Women* (New York: Penguin, 2012).

53 **nearly one-third of all married women** Paul R. Amato et al., *Alone Together: How Marriage in America Is Changing* (Cambridge, MA: Harvard University Press, 2009), 150.

53 **men and women work roughly the same number of hours** Rachel Krantz-Kent, "Measuring Time Spent in Unpaid Household Work: Results from the American Time Use Survey," *Monthly Labor Review* 132, no. 7 (2009): 46–59.

53 **2011 cover story called "Chore Wars"** Ruth D. Konigsberg, "Chore Wars," *Time,* August 8, 2011, 44.

54 **"I do the upstairs, Evan does the downstairs"** Hochschild, *The Second Shift,* 46.

54 **"When couples struggle"** Ibid., 19.

54 **"The deeper problem such women face"** Ibid., 273.

54 **"household division of labor being a key source"** Amato et al., *Alone Together,* 153–54, 156.

54 **perhaps the most intriguing tidbit** Darby Saxbe and Rena L. Repetti, "For Better or Worse? Coregulation of Couples' Cortisol Levels and Mood States," *Journal of Personality and Social Psychology* 98, no. 1 (2010): 92–103.

55 **mothers of children under six still work five hours more per week** Konigsberg, "Chore Wars," 48.

55 **women were three times more likely than men** Sarah A. Burgard, "The Needs of Others: Gender and Sleep Interruptions for Caregiving," *Social Forces* 89, no. 4 (2011): 1189–1215.

55 **I once sat on a panel with Adam Mansbach** Brooklyn Book Festival, "Politically Incorrect Parenting," panel discussion, September 18, 2011.

56 **"satisfaction with the division of the child-care tasks"** Cowan and Cowan, *When Partners Become Parents,* 142.

56 **women . . . still devote nearly twice as much time** Bianchi, "Family Change," 27, 29.

56 **a father in a room by himself** Belinda Campos et al., "Opportunity for Interaction? A Naturalistic Observation Study of Dual-Earner Families After Work and School," *Journal of Family Psychology* 23, no. 6 (2009): 798–807.

57 **if a married mother believes that child care is unfairly divided** Amato et al., *Alone Together,* 170. Additionally, in 2012 the Organization for Economic Cooperation and Development (OECD) determined that women in the United States work twenty-one minutes more per day than men, which is exactly on a par with the worldwide average. Catherine Rampell, "In Most Rich Countries, Women Work More Than Men," Economix (blog), *New York Times,* December 19, 2012, available at: http://economix.blogs.nytimes.com/2012/12/19/in-most-rich-countries-women-work-more-than-men.

57 **a father . . . is more apt to get involved in "interactive" tasks** Suzanne M. Bianchi, John P. Robinson, and Melissa A. Milkie, *Changing Rhythms of American Family Life* (New York: Russell Sage Foundation, 2006), 66–67.

57 **fathers guessed that they did . . . about 42 percent of the child care** Amato et al., *Alone Together,* 150.

57 **The actual number that year** Bianchi, "Family Change," 7, 9.

58 **it remains roughly the same today** Kim Parker and Wendy Wang, "Modern Parenthood: Roles of Moms and Dads Converge as They Balance Work and Family," *Pew Research Social & Demographic Trends*, March 14, 2013, available at http://www.pewsocialtrends.org/2013/03/14/modern-parenthood-roles-of-moms-and-dads-converge-as-they-balance-work-and-family/.

58 **42 percent of married fathers reported multitasking "most of the time"** Bianchi et al., *Changing Rhythms,* 136 (chart).

58 **two sociologists provided an even more granular analysis** Shira Offer and Barbara Schneider, "Revisiting the Gender Gap in Time-Use Patterns: Multitasking and Well-being Among Mothers and Fathers in Dual-Earner Families," *American Sociological Review* 76, no. 6 (2011): 809–33.

59 **women are more likely than men to feel "always rushed"** Marybeth J. Mattingly and Liana C. Sayer, "Under Pressure: Gender Differences in the Relationship Between Free Time and Feeling Rushed," *Journal of Marriage and Family* 68 (2006): 205–21.

60 **fall disproportionately to them** Ibid., 216.

60 **she ought to ask them to devise a pie chart of their identities** Cowan and Cowan, *When Partners Become Parents,* 82.

60 **mothers who carry the child in lesbian couples** Charlotte J. Patterson, "Families of the Lesbian Baby Boom: Parents' Division of Labor and Children's Adjustment," *Developmental Psychology* 31, no. 1 (1995): 115-123.

60 **What *did* surprise the Cowans** Much of the background information on Philip and Carolyn Cowan's research is taken from a series of interviews with the author, February 2, 2011, and March 10, 2011.

60 **the greater the disparity between how a mother and a father** Cowan and Cowan, *When Partners Become Parents,* 81.

61 **80 percent of them believed** Mom Central, "How Moms Socialize Online – Part 1," *Revolution + Research = R2* (blog), December 1, 2010, http://www.momcentral.com/blogs/revolution-research-r2/how-moms-socialize-online-part-1.

61 **women's social networks—and the frequency of their contact** Allison Munch, J. Miller McPherson, and Lynn Smith-Lovin, "Gender, Children, and Social Contact: The Effects of Childrearing for Men and Women," *American Sociological Review* 62 (1997): 509-520.

61 **"That really surprised me"** Kathryn Fink, interview with the author, February 24, 2012.

62 **Sociologists who have examined** Masako Ishii-Kuntz and Karen Seccombe, "The Impact of Children upon Social Support Networks throughout the Life Course," *Journal of Marriage and the Family* 51 (1989): 777–790, especially Table 3 on page 783.

62 **"machers" and "schmoozers"** Robert Putnam, *Bowling Alone: The Collapse and Revival of American Community* (New York: Touchstone, 2000), 93.

62 **"Holding other demographic features constant"** Ibid., 278.

63 **"Women who have worked for a number of years"** Benjamin Spock, *Problems of Parents* (Boston: Houghton Mifflin, 1962), 34.

64 **"discuss important matters"** Miller McPherson, Lynn Smith-Lovin, and Matthew E. Brashears, "Social Isolation in America: Changes in Core Discussion Networks over Two Decades," *American Sociological Review* 71 (2006): 353-375.

64 **the decline of almost every measurable form of civic participation** Putnam, "Civic Participation," in *Bowling Alone,* 48.

64 **number of times married Americans spent a social evening** Ibid., 105.

65 **subsequent studies have shown that this number continued to drop through 2008** Peter V. Marsden, ed., *Social Trends in American Life: Finding from the General Social Survey since 1972* (Princeton: Princeton University Press, 2012), 244.

65 **"pervasive busyness"** Putnam, *Bowling Alone,* 189.

65 **In the mid- to late seventies** Ibid., 98.

66 **"extended families have never been the norm in America"** Coontz, *The Way We Never Were,* 12.

66 **college-educated Americans tend to live farther away from their parents** Janice Compton and Robert A. Pollak, "Proximity and Coresidence of Adult Children and Their Parents: Description and Correlates" (working paper), Ann Arbor: University of Michigan, Retirement Research Center (October 2009).

67 **the so-called sandwich generation** George James, "A Survival Course for the Sandwich Generation," *New York Times,* January 17, 1999. See also Carol Abaya's website (http://www.thesandwichgeneration.com/index.htm); Abaya, the subject of the *Times* article, holds a registered trademark on the term "sandwich generation."

69 **data about compliance requests** Gerald R. Patterson, "Mothers: The Unacknowledged Victims," *Monographs of the Society for Research in Child Development* 45, no. 5 (1980): 1–64.

69 **researchers from Emory and the University of Georgia** Rex Forehand et al., "Mother-Child Interactions: Comparison of a Non-Compliant Clinic Group and a Non-Clinic Group," *Behaviour Research and Therapy* 13 (1975): 79–84.

69 **all the way to the present day** See, for example, Leon Kuczynski and Grazyna Kochanska, "Function and Content of Maternal Demands: Developmental Significance of Early Demands for Competent Action," *Child Development* 66 (1995): 616–28; Grazyna Kochanska and Nazan Aksan, "Mother-Child Mutually Positive Affect, the Quality of Child Compliance to Requests and Prohibitions, and Maternal Control as Correlates of Early Internalization," *Child Development* 66, no. 1 (1995): 236–54.

69 **averaging a conflict every two and a half minutes** Margaret O'Brien Caughy, Keng-Yen Huang, and Julie Lima, "Patterns of Conflict Interaction in Mother-Toddler Dyads: Differences Between Depressed and Non-depressed Mothers," *Journal of Child and Family Studies* 18 (2009): 10–20.

69 **"Much of contemporary developmental psychology"** Urie Bronfenbrenner, *The Ecology of Human Development: Experiments by Nature and Design* (Cambridge, MA: Harvard University Press, 1979), 18.

71 **Before the late eighteenth century** Amato et al., *Alone Together,* 12–13.

72 **94 percent of singles** David Popenoe and Barbara Defoe Whitehead, eds.,

"The State of Our Unions: 2001," National Marriage Project, available at: http://www.stateofourunions.org/pdfs/SOOU2001.pdf (accessed March 30, 2013).

72 **"SuperRelationship"** Ibid.

72 **couple-time drops by two-thirds** Bianchi et al., *Changing Rhythms,* 104.

72 **the story of a beautiful couple** William Doherty, interview with the author, January 26, 2011.

73 **having sex less than once a week** R. Kumar, H. A. Brant, and Kay Mordecai Robson, "Childbearing and Maternal Sexuality: A Prospective Survey of 119 Primiparae," *Journal of Psychosomatic Research* 25, no. 5 (1981): 373–83.

73 **"jobs, commuting, housework"** Cathy Stein Greenblat, "The Salience of Sexuality in the Early Years of Marriage," *Journal of Marriage and Family* 45, no. 2 (1983): 289–99.

73 **the presence of young children** Vaughn Call, Susan Sprecher, and Pepper Schwartz, "The Incidence and Frequency of Marital Sex in a National Sample," *Journal of Marriage and Family* 57, no. 3 (1995): 639–52.

73 **"Respondents with low and high educational attainment levels"** Ibid.

74 **the most precipitous drop** Ibid.

75 **"In our erotic lives we abandon our children"** Adam Phillips, *Side Effects* (New York: HarperCollins, 2006), 73–74.

76 **"there were no differences between homemakers and women employed"** Janet Shibley Hyde, John D. DeLamater, and Amanda M. Durik, "Sexuality and the Dual-Earner Couple, Part II: Beyond the Baby Years," *Journal of Sex Research* 38, no. 1 (2001): 10–23.

79 **"She knows something is up"** Michael Cunningham, *A Home at the End of the World* (New York: Picador, 1990), 26.

82 **the difference in happiness levels between parents and nonparents** Robin W. Simon, Jennifer Glass, and M. Anders Anderson, "The Impact of Parenthood on Emotional and Physical Well-being: Some Findings from a Cross-National Study," paper presented at the Thirteenth International Conference of Social Stress Research, Dublin, Ireland (June 22, 2012).

82 **"the happiness that people derive from parenthood"** Arnstein Aassve, Letizia Mencarini, and Maria Sironi, "Institutional Transition, Subjective Well-being, and Fertility," paper presented at the 2013 annual meeting of the Population Association of America, New Orleans, LA (April 11, 2013).

82 **"These countries . . . are scoring on a whole range of categories"** Arnstein Aassve, interview with the author, April 9, 2013.

83 **"My elder daughter, from the time she was eighteen months of age"** Warner, *Perfect Madness*, 10.

83 **it cost more for families to put two children in day care** Child Care Aware of America, "Parents and the High Cost of Child Care, 2012 Report," avail-

able at: http://www.naccrra.org/sites/default/files/default_site_pages/2012/cost_report_2012_final_081012_0.pdf (accessed May 5, 2013).

83 **compared the moment-to-moment well-being of women** Daniel Kahneman et al., "The Structure of Well-being in Two Cities: Life Satisfaction and Experienced Happiness in Columbus, Ohio, and Rennes, France," in *International Differences in Well-being*, ed. Ed Diener, Daniel Kahneman, and John Helliwell (Oxford: Oxford University Press, 2010).

84 **"spend less of the afternoon driving children"** Daniel Kahneman, *Thinking, Fast and Slow* (New York: Farrar, Straus and Giroux, 2011), 394.

85 **feel like they don't have enough time for themselves** Bianchi et al., *Changing Rhythms*, 135.

85 **"unentitlement"** Cowan and Cowan, *When Partners Become Parents*, 196.

89 **"you often get all these attributions"** Philip and Carolyn Cowan, interviews with the author, February 2, 2011, and March 10, 2011.

91 **all it takes for a couple to start fighting** Michael Lewis, *Home Game: An Accidental Guide to Fatherhood* (New York: W. W. Norton, 2009), 11, 13.

91 **solution to these excesses is to emulate the French** Druckerman, *Bringing up Bébé*.

91 **"consumer parenting"** See, for example, William J. Doherty, *Take Back Your Marriage: Sticking Together in a World That Pulls Us Apart* (New York: Guilford Press, 2001), 53.

93 **couples who had hashed out divisions of labor during pregnancy** Cowan and Cowan, *When Partners Become Parents*, 176.

chapter three

95 **"He loves to see his son's wit"** Michael Ondaatje, *The English Patient* (New York: Vintage Books, 1992), 301.

98 **"There is a certain part of all of us"** Milan Kundera, *Immortality*, trans. Peter Kussi (New York: HarperCollins, 1990), 4.

99 **"boundless and unwearied in giving"** C. S. Lewis, *The Four Loves* (Boston: Houghton Mifflin Harcourt, 1991), 8.

101 **"this childish uninhibitedness"** Gopnik, *The Philosophical Baby*, 72.

102 **"gnashed their terrible teeth"** Maurice Sendak, *Where the Wild Things Are* (New York: HarperCollins, 1988).

102 **"writers as diverse as Wordsworth and Freud"** Phillips, *Going Sane*, 92.

103 **He quotes the analyst Donald Winnicott** Ibid., 81.

103 **"mad in the best sense of the word"** Ibid.

103 **"For Winnicott the question was not"** Ibid., 79.

104 **"despite the proliferation of contrived metrics"** Matthew B. Crawford,

Shop Class as Soulcraft: An Inquiry into the Value of Work (New York: Penguin, 2009), 8.

105 **the overwhelming majority of Americans** Harris Interactive Poll, "Three in Ten Americans Love to Cook, While One in Five Do Not Enjoy It or Don't Cook," July 27, 2010, available at: http://www.harrisinteractive.com/vault/HI-Harris-Poll-Cooking-Habits-2010-07-27.pdf (accessed on April 10, 2013).

105 **"the *experience* of making things and fixing things"** Crawford, *Shop Class as Soulcraft,* 3–4.

105 **"things" and "devices"** Ibid., 65–66.

106 **"*inherently* instrumental, or pragmatically oriented"** Ibid., 68.

106 **the minds of babies and young children** Gopnik, *The Philosophical Baby,* 157–58.

107 **the time he took one of his sons to the beach** Mihaly Csikszentmihalyi, interview with the author, July 25, 2011.

108 **asking pointless questions is the true specialty of children** Gareth B. Matthews, *The Philosophy of Childhood* (Cambridge, MA: Harvard University Press, 1996), 5.

108 **"It sharpens the mind by narrowing it"** Quoted in Oliver Wendell Holmes, "Brown University—Commencement 1897," in *Collected Legal Papers,* ed. Harold J. Laski (New York: Harcourt, Brace, and Howe, 1920), 164.

108 **"That is hard for adults"** Matthews, *Philosophy of Childhood,* 18.

108 **"What, then, is time?"** Quoted in ibid., 13.

109 **"Papa, how can we be sure that everything is not a dream?"** Ibid., 17.

109 **asked his mother on the car ride home** Ibid., 28.

109 **"philosophy is an adult attempt to deal"** Ibid., 13.

109 **"If it cannot *answer* so many questions"** Quoted in Gareth B. Matthews, *Philosophy and the Young Child* (Cambridge, MA: Harvard University Press, 1980), 2.

111 **Gift-love and Need-love** Lewis, *The Four Loves,* 1.

112 **"we love them because we care for them"** Gopnik, *The Philosophical Baby,* 243.

113 **"There is something in each of us"** Lewis, *The Four Loves,* 133, 135.

114 **"There are many ways to approach that ideal"** Gopnik, *The Philosophical Baby,* 243.

chapter four

117 **"Profound must be the depths of the affection"** Edward S. Martin, *The Luxury of Children and Some Other Luxuries* (New York: Harper & Brothers, 1904), 135.

119 **"overscheduled kids"** Doherty interview. See also William Doherty and Barbara Z. Carlson, "Overscheduled Kids and Underconnected Families," in *Take Back Your Time: Fighting Overwork and Time Famine in Families*, ed. J. de Graaf (San Francisco: Berritt Koehler, 2003), 38-45.

120 **"accomplishment of natural growth"** Annette Lareau, *Unequal Childhoods: Class, Race, and Family Life* (Berkeley: University of California Press, 2003), 3.

120 **"Concerted cultivation . . . places intense labor demands"** Ibid., 13.

120 **"Unlike in working-class and poor families"** Ibid., 171, 175.

122 **mothers *still* spent 3.7 fewer hours per week** Bianchi, "Family Change," 27, 29.

126 **"a time of deficiency and incompleteness"** Steven Mintz, *Huck's Raft: A History of American Childhood* (Cambridge, MA: Harvard University Press, 2004), 3.

126 **it was not uncommon for . . . colonists to call their newborns "it"** Zelizer, *Pricing the Priceless Child*, 25.

126 **"Children suffered burns"** Mintz, *Huck's Raft*, 17, 20.

127 **Americans hardly started with the notion that children** Ibid., 3, 77, 80, 90.

127 **children *already* were integral to the farm economy** Ibid., 135.

127 **children were more likely to earn money for their families** Zelizer, *Pricing the Priceless Child*, 59.

127 **the wages of teenage boys often exceeded those of their dads** Mintz, *Huck's Raft*, 136.

128 **Childhood as we think of it today** Ibid., 3.

128 **that strange custom we all know as "the allowance"** Zelizer, *Pricing the Priceless Child*, 104.

128 **"The useful labor of the nineteenth century"** Ibid., 97, 98.

128 **"economically worthless but emotionally priceless"** Ibid., 14.

128 **a "filiarchy," or culture in which kids run the show** William H. Whyte, "How the New Suburbia Socializes," *Fortune* (August, 1953), 120.

129 **"Middle-class children . . . argue with their parents"** Lareau, *Unequal Childhoods*, 13, 153.

129 **"The very same skills parents encourage in their children"** Ibid., 111.

131 **In 1990 Sugar Land was 79 percent white** For Sugar Land demographics, see US Census Bureau, "State and County QuickFacts, Sugar Land, Texas," available at: http://quickfacts.census.gov/qfd/states/48/4870808.html (accessed April 19, 2013).

133 **the "Top 10 Percent Rule"** Texas House bill 588 (1997).

133 **in fact, they make up just 31 percent** Mary Lou Robertson, Fort Bend Independent School District, communication with the author, May 18, 2012.

134 **Duke University Talent Identification Program** For information on TIP, see the Duke University website, http://www.tip.duke.edu (accessed April 19, 2013).

135 **"In other societies, where parents were bringing up children"** Margaret Mead, *And Keep Your Powder Dry: An Anthropologist Looks at America* (New York: Berghahn Books, 2000), 63.

135 **"The American parent expects his child to leave him"** Ibid., 24.

136 **"autumnal"** Ibid., 28.

136 **"We find new schools of education"** Ibid., 64, 65.

137 **"all one can do is make him strong"** Ibid., 25.

137 **"that all young people should follow"** Mintz, *Huck's Raft,* 383.

138 **"Parenting [is] not simply about raising a child"** Nora Ephron, *I Feel Bad About My Neck: And Other Thoughts on Being a Woman* (New York: Vintage, 2006), 58.

138 **the Immigration and Nationality Act of 1965** Immigration and Nationality Act, P.L. 89-236, 79 Stat. 911 (1965).

139 **more African American . . . and not quite as well-to-do** US Census Bureau, "American FactFinder," available at: http://factfinder2.census.gov/faces/nav/jsf/pages/index.xhtml (accessed April 21, 2013).

143 **the average household debt exceeded its disposable income by 34 percent** Josh Sanburn, "Household Debt Has Fallen to 2006 Levels, But Not Because We've Grown More Frugal," Economy (blog), *Time,* October 19, 2012, available at http://business.time.com/2012/10/19/household-debt-has-fallen-to-2006-levels-but-not-because-were-more-frugal/.

143 **a record number of American households** Warner, *Perfect Madness,* 201–2.

143 **9 percent of their income** Office of the Vice President of the United States, Middle Class Task Force, "Why Middle Class Americans Need Health Reform," available at: http://www.whitehouse.gov/assets/documents/071009_FINAL_Middle_Class_Task_Force_report2.pdf (accessed April 22, 2013).

143 **Between 1980 and 2009,** Frank Levy and Thomas Kochan, "Addressing the Problem of Stagnant Wages," Employment Policy Research Network, available at: http://www.employmentpolicy.org/sites/www.employmentpolicy.org/files/field-content-file/pdf/Mike%20Lillich/EPRN%20WagesMay%2020%20-%20FL%20Edits_0.pdf (accessed April 22, 2013).

143 **the wage gap between mothers and childless women** "The Motherhood Penalty: Stanford Professor Shelley Correll," Clayman Institute, available at: http://www.youtube.com/watch?v=vLB7Q3_vgMk (accessed April 22, 2013).

143 **which estimates that a child born in 2010** US Department of Agriculture,

"Expenditures on Children by Families, 2010," ed. Mark Lino, available at: http://www.cnpp.usda.gov/publications/crc/crc2010.pdf.

143 **These price tags do not include college tuition** US Department of Education, National Center for Education Statistics, *Digest of Education Statistics: 2011* (2012): table 349.

144 **the best-paid men . . . are far more apt to put in long** Peter Kuhn and Fernando Lozano, "The Expanding Workweek? Understanding Trends in Long Work Hours Among US Men, 1979–2004," *Journal of Labor Economics* (December 2005): 311–43.

149 **"women's lives are much more heavily intertwined"** Annette Lareau and Elliot B. Weininger, "Time, Work, and Family Life: Reconceptualizing Gendered Time Patterns Through the Case of Children's Organized Activities," *Sociological Forum* 23, no. 3 (2008): 422, 427.

150 **It was mothers who signed their children up** Ibid., 427.

150 **"at least some employed mothers face a tradeoff"** Ibid., abstract.

150 **"pressure points"** Ibid., 422, 442.

150 **"women should return"** "Partisan Polarization Surges in Bush, Obama Years." *Pew Research Social & Demographic Trends,* June 4, 2012, available at http://www.people-press.org/2012/06/04/section-2-demographics-and-american-values/.

150 **Children would be better off** Wendy Wang, Kim Parker, and Paul Taylor, "Breadwinner Moms," *Pew Research Social & Demographic Trends,* May 29, 2013, available at http://www.pewsocialtrends.org/2013/05/29/breadwinner-moms/.

150 **whenever the free market threatens** Sharon Hays, *The Cultural Contradictions of Motherhood* (New Haven, CT: Yale University Press, 1996).

151 **"in the workplace, a woman . . . must be efficient"** T. Berry Brazelton, *Working and Caring* (New York: Perseus, 1987), xix.

152 **Thanks to his mother's college education** Hulbert, *Raising America,* 32.

152 **"By some strange cosmic alchemy"** Quoted in ibid., 101.

152 **Even the year the word "parent" first gained popularity** Ibid., 281.

153 **The average age of first marriage** Friedan, *The Feminine Mystique,* 243.

153 **"Occupation: Housewife"** This term appears throughout *The Feminine Mystique*: pp. 44, 61, 89, 91–93, 103, 118, 298, 334, 350, 435, 461, 488.

153 **"the problem that has no name"** Ibid., 57.

153 **"One of the ways that the housewife raises her own prestige"** Ibid., 310.

154 **17.5 hours per week** Bianchi, "Family Change," 27.

155 **"A good woman would have given over her life"** Erica Jong, *Fear of Flying* (New York: NAL Trade, 2003), 210. (And a hat-tip to Claire Dederer's *Poser,* without which I never would have remembered this perfect quote.)

155 **the title of a 2009 book of essays** Ayelet Waldman, *Bad Mother: A Chronicle of Maternal Crimes, Minor Calamities, and Occasional Moments of Grace* (New York: Doubleday, 2009).

155 **"intensive mothering"** Sharon Hays, *The Cultural Contradictions of Motherhood* (New Haven, CT: Yale University Press, 1996), 4.

155 **"The vast majority of these women"** Ibid., 146.

158 **less than one-fourth the income of two-parent families** Bianchi, "Family Change," 31.

158 **they have more health problems** Roni Caryn Rabin, "Disparities: Health Risks Seen for Single Mothers," *New York Times*, June 13, 2011.

158 **fewer social ties** Jennifer A. Johnson and Julie A. Honnold, "Impact of Social Capital on Employment and Marriage Among Low Income Single Mothers," *Journal of Sociology and Social Welfare* 38, no. 4 (2011): 11.

158 **more likely to receive child support** Bianchi, "Family Change," 31.

158 **their children spend more time under their roofs** Linda Nielsen, "Shared Parenting After Divorce: A Review of Shared Residential Parenting Research," *Journal of Divorce and Remarriage* 52 (2011): 588.

158 **"have as many demands on their time"** Bianchi, "Family Change," 30.

159 **report multitasking "most of the time"** Ibid., 106.

159 **fewer hours . . . socializing . . . having meals** Ibid., 96.

159 **"the mommy mystique"** Warner, *Perfect Madness*, ch. 1.

159 **fathers work longer hours** Aumann et al., "The New Male Mystique," 11.

160 **work-family conflict** Ibid., 2.

160 **84 percent of male respondents believed** Ibid., 7.

160 **getting office messages during non-office hours** Ibid., 6.

160 **"At my job I have to work very hard"** Ibid.

160 **"They don't want to be stick figures in their children's lives"** Ellen Galinsky, interview with the author, April 29, 2010.

164 **"Some historians even maintain"** Howard Chudacoff, *Children at Play: An American History* (New York: New York University Press, 2007), 6.

164 **a room of their own** Quoted in Zelizer, *Pricing the Priceless Child*, 53–54.

165 **toy sales . . . had reached $1.25 billion** Mintz, *Huck's Raft*, 277.

165 **"for babies between birth and age two alone"** Pamela Paul, *Parenting, Inc.* (New York: Times Books, 2008), 10.

165 **domestic sales of kids' toys** Toy Industry Association, "Annual Sales Data," available at: http://www.toyassociation.org (accessed April 23, 2013).

165 **"Modern manufactured toys implied a solitariness"** Mintz, *Huck's Raft*, 217.

165 **Crayons . . . Tinker Toys . . . Lincoln Logs . . . Legos** Chudacoff, *Children at Play*, 118–19.

165 **"one defining feature of young people's lives"** Mintz, *Huck's Raft,* 347.

165 **22 percent of American children today are only-children** Rose M. Kreider and Renee Ellis, "Living Arrangements of Children: 2009," *Current Population Reports*, U.S. Census Bureau (2011): 70–126.

165 **"Middle-class parents worry"** Lareau, *Unequal Childhoods,* 185.

165 **"we were bored all the time"** Nancy Darling, "Are Today's Kids Programmed for Boredom?" Thinking About Kids (blog), *Psychology Today,* November 30, 2011, available at: http://www.psychologytoday.com/blog/thinking-about-kids/201111/are-todays-kids-programmed-boredom.

167 **"kids have very little experience"** Ibid.

167 **"No kid ever says"** Nancy Darling, interview with the author, March 3, 2011.

167 **"Middle-class children often feel *entitled*"** Lareau, *Unequal Childhoods,* 81.

167 **there were fewer than 100 playgrounds** Mintz, *Huck's Raft,* 179.

167 **In 1906** Ibid.

168 **the number of elementary- and middle-school students who walk or bike to school** Noreen C. McDonald, Austin L. Brown, Lauren M. Marchetti, and Margo S. Pedroso, "U.S. School Travel, 2009: An Assessment of Trends," *American Journal of Preventative Medicine*, 41(2) (2011): 148.

168 **crimes against children have been steadily declining** David Finkelhor, Lisa Jones, and Anne Shattuck, "Updated Trends in Child Mistreatment, 2011," Crimes Against Children Research Center, University of New Hampshire (January 2013), available at http://www.unh.edu/ccrc/pdf/CV203_Updated%20trends%202011_FINAL_1-9-13.pdf.

168 **reports of child sexual abuse fell by 63 percent** Ibid.

168 **"In retrospect, one can see how terrified parents"** Mintz, *Huck's Raft,* 336.

168 **wave of alarm over stranger abductions and madmen inserting razor blades** Ibid.

168 **popping up on milk cartons . . . number of abductions by strangers** Ibid., 337. The 1-in-115,000 figure is derived from applying Mintz's statistic to the U.S. Census Bureau's figure for the total number of children under 18 in the United States.

168 **four times that many died** National Highway Traffic Safety Administration, "Fatality Analysis Reporting System," available at: http://www-fars.nhtsa.dot.gov (accessed April 22, 2013).

169 **have assaulted people they know** Howard N. Snyder, "Sexual Assault of Young Children as Reported to Law Enforcement: Victim, Incident, and Offender Characteristics" (Washington, DC: US Department of Justice, Bureau of Justice Statistics, July 2000), 10.

170 **eight- to ten-year-olds play them** Victoria J. Rideout, Ulla G. Foehr, and Donald F. Roberts, "Generation M^2: Media in the Lives of Eight- to Eighteen-Year-Olds," *Kaiser Family Foundation* (January 2010): 5, 15.

170 **Sixty percent of all "heavy" media users** Ibid., 5.

170 **63 percent of seventh- and eighth-grade boys** Lawrence Kutner and Cheryl Olson, *Grand Theft Childhood: The Surprising Truth About Violent Video Games, and What Parents Can Do* (New York: Simon & Schuster, 2008), 90.

170 **young men were degenerating into dandies** Mintz, *Huck's Raft,* 193.

171 **How to tie five essential knots** Conn Iggulden and Hal Iggulden, *The Dangerous Book for Boys* (New York: HarperCollins, 2007), passim.

171 **"There's now this weird structured tension"** Mimi Ito, interview with the author, May 24, 2012.

174 **"sacralization" of childhood** Zelizer, *Pricing the Priceless Child,* 22.

174 **"Relieved of having to carry out"** Quoted in Hulbert, *Raising America,* 101.

175 **"Individual happiness becomes that elusive good"** Hays, *The Cultural Contradictions of Motherhood,* 67.

175 **"It is unrealistic, I think"** Phillips, *On Balance,* 90.

176 **"points to a plowed field"** Jerome Kagan, "The Child in the Family," *Daedalus* 106, no. 2 (1977): 33–56; for a thorough overview of Kagan's study, see Zelizer, *Pricing the Priceless Child,* 220.

176 **"We are uncertain about how we want our children to behave"** Spock, *Problems of Parents,* 290.

176 **even Chua has questions about this approach** Amy Chua website, "From Author Amy Chua," available at: http://amychua.com (accessed April 22, 2013).

179 **Suzuki, a method of musical instruction first developed** Talent Education Research Institute, "Suzuki Method," available at: http://www.suzukimethod.or.jp/indexE.html accessed April 22, 2013).

179 **whose numbers have fallen quite a bit** Putnam, *Bowling Alone,* 100. For more updated information about declines in family dinners, see Robert Putnam's forthcoming book in 2014.

180 **couples spent . . . 12.4 hours alone together per week** Bianchi et al., *Changing Rhythms,* 104.

180 **those kinds of volunteer efforts and public involvements** Putnam, *Bowling Alone,* ch. 7.

180 **how one raises a child** Kagan, "Our Babies, Our Selves," 42.

181 **The data were quite clear** Ellen Galinsky, *Ask the Children: What America's Children Really Think About Working Parents* (New York: William Morrow, 1999), xv.

chapter five

183 **"They don't tell you"** Dani Shapiro, *Family History: A Novel* (New York: Anchor Books, 2004), 120.

185 **"getting wenches with child"** *The Winter's Tale,* ed. Jonathan Bate and Eric Rasumssen (New York: Modern Library, 2009), act 3, scene 3, lines 64–65.

185 **"so that someone is happy to see you"** Ephron, *I Feel Bad About My Neck,* 125.

186 **"It doesn't seem to me like adolescence is a difficult time"** Laurence Steinberg, interview with the author, April 11, 2011.

186 **"The hormonal changes of puberty"** Laurence Steinberg, *Adolescence,* 10th ed. (New York: McGraw-Hill, 2014), 418.

186 **longitudinal study he conducted of over two hundred families** Laurence Steinberg, *Crossing Paths: How Your Child's Adolescence Triggers Your Own Crisis* (New York: Simon & Schuster, 1994), 17, 253, 254–55.

186 **feelings of rejection . . . symptoms of distress** Ibid., 28.

187 **"We were much better able to predict"** Ibid., 59.

187 **It was "discovered" by Stanley Hall** See, for example, Jeffrey Jensen Arnett, "G. Stanley Hall's *Adolescence*: Brilliance and Nonsense," *History of Psychology* 9, no. 3 (2006): 186–97.

189 **"We were astounded by the enthusiastic response"** Steinberg, *Crossing Paths,* 17.

193 **the proportion of waking hours that children spent with their families** Reed W. Larson et al., "Changes in Adolescents' Daily Interactions with Their Families from Ages 10 to 18: Disengagement and Transformation," *Developmental Psychology* 32, no. 4 (1996): 752.

193 **"During childhood, it's about trying to help develop"** Joanne Davila, interview with the author, April 8, 2011.

194 **"somebody who is trying to get himself kidnapped"** Phillips, *On Balance,* 102.

194 **"being involved in their children's education"** Galinsky, *Ask the Children,* 45.

194 **"How sharper than a serpent's tooth"** William Shakespeare, *King Lear,* 2d ed., ed. Elspeth Bain et al. (Cambridge: Cambridge University Press, 2009), act 1, scene 4, lines 243–44.

195 **"second only to infancy"** Gerald Adams and Michael Berzonsky, eds., *Blackwell Handbook of Adolescence* (Malden, MA: Blackwell Publishing, 2006), 66.

195 **"old script"** Steinberg, *Crossing Paths,* 209.

195 **"I believe that we have underestimated"** Ibid., 62.

196 **precisely the conclusion of a 1998 meta-study** Laursen et al., "Reconsidering Changes in Parent-Child Conflict."

196 **In her work, Nancy Darling offers a nuanced analysis** For example, see Nancy Darling, Patricio Cumsille, and M. Loreto Martinez, "Individual Differences in Adolescents' Beliefs About the Legitimacy of Parental Authority and Their Own Obligation to Obey: A Longitudinal Investigation," *Child Development* 79, no. 4 (2008): 1103–118.

196 **Most kids . . . have no objections** Nancy Darling, interview with the author, March 29, 2011.

197 **wearing jeans to church** Nancy Darling, "The Language of Parenting: Legitimacy of Parental Authority," Thinking About Kids (blog), *Psychology Today* (January 11, 2010), available at: http://www.psychologytoday.com/blog thinking-about-kids/201001/the-language-parenting-legitimacy-parental-authority.

198 **One is being divorced** Steinberg, *Crossing Paths*, 234, 237.

198 **parents tend to be much *closer*** Steinberg, *Crossing Paths,* 233.

198 **"I think it's a lot easier to parent a child"** Brené Brown, interview with the author, September 18, 2012.

199 **"The critical protective variable"** Steinberg, *Crossing Paths,* 239.

203 **sexual frequency among married couples declines** See, for example, Call et al., "The Incidence and Frequency of Marital Sex."

203 **"growth spurt[s], growth of body hair, and skin changes"** Shawn D. Whiteman, Susan M. McHale, and Ann C. Crouter, "Longitudinal Changes in Marital Relationships: The Role of Offspring's Pubertal Development," *Journal of Marriage and* Family 69, no. 4 (2007): 1009.

203 **"They have weathered a lot of storms"** Thomas Bradbury, communication with the author, August 15, 2012.

204 **"Inevitably, we see ourselves in our kids"** Andrew Christensen, interview with the author, May 18, 2011.

205 **In one intriguing study** Christy M. Buchanan and Robyn Waizenhofer, "The Impact of Interparental Conflict on Adolescent Children: Considerations of Family Systems and Family Structure," in *Couples in Conflict,* ed. Alan Booth, Ann C. Crouter, and Mari Clements (Mahwah, NJ: Lawrence Erlbaum Associates, 2001), 156.

205 **"In fact, the more frequently the teenager dated"** Steinberg, *Crossing Paths,* 178–79.

205 **"At least there are coaches for breast-feeding"** Susan McHale, interview with the author, September 12, 2012.

205 **"One parent is the softie"** Andrew Christensen, interview with the author, May 18, 2011.

207 **a large, renowned longitudinal study by the University of Michigan** I am deeply indebted to U. J. Moon of the Maryland Population Research Center for her original synthesis of data provided by the University of Michigan study. The raw data from which U.J. derived her numbers can be found in the Panel Study of Income Dynamics, 2002 public use dataset, produced and distributed by the Institute for Social Research, Survey Research Center, University of Michigan (2012).

207 **adolescent girls and boys *both* direct** Darling et al., "Aggression During Conflict." See also Nancy Darling et al., "Within-Family Conflict Behaviors as Predictors of Conflict in Adolescent Romantic Relations," *Journal of Adolescence* 31 (2008): 671–90.

207 **mothers are also more likely than fathers to quarrel** Steinberg, *Crossing Paths,* 200.

207 **bring more family stress into their workplaces** Ibid., 200.

207 **"Women's personal crises at midlife"** Ibid., 256.

209 **adolescents *overestimate* risk** For a technical perspective on this, see Wändi Bruine de Bruin, Andrew M. Parker, and Baruch Fischoff, "Can Adolescents Predict Significant Life Events?" *Journal of Adolescent Health* 41 (2007): 208–10. For a layperson's perspective, see David Dobbs, "Teenage Brains," *National Geographic* 220, no. 4 (October 2011): 36–59.

210 **the prefrontal cortex** For a thorough, uncomplicated review of how the adolescent brain works and evolves, see Daniel R. Weinberger, Brita Elvevag, and Jay N. Giedd, "The Adolescent Brain: A Work in Progress," report of the National Campaign to Prevent Teen Pregnancy (June 2005).

210 **a huge flurry of activity** Sarah-Jayne Blakemore and Suparna Choudury, "Brain Development During Puberty: State of the Science" (commentary), *Developmental Science* 9, no. 1 (2006): 11–14.

210 **why adolescents seem so fond of arguing** Nancy Darling, "What Middle School Parents Should Know Part 2: Adolescents Are Like Lawyers," Thinking About Kids (blog), *Psychology Today* (September 9, 2010), available at: http://www.psychologytoday.com/blog/thinking-about-kids/201009/what-middle-school-parents-should-know-part-2-adolescents-are-lawyer.

210 **their prefrontal cortexes are still adding myelin** Weinberger et al., "The Adolescent Brain," 9–10.

210 **"They're kind of flying by the seat of their pants"** B. J. Casey, interview with the author, August 28, 2012.

211 **"Teenagers are more Kirk than Spock"** Ibid.

211 **"And then parents are going to get into tussles"** Laurence Steinberg, interview with the author, April 11, 2011.

211 **adolescent brains are also more prone to substance abuse and dependence**

Linda Patia Spear, "Alcohol's Effects on Adolescents" (sidebar), *Alcohol Research and Health* 26, no. 4 (2002): 288.

211 **"I used to think that if I locked up my son"** Casey interview.

211 **victor in the quest for control** Siobhan S. Pattwell et al., "Altered Fear Learning Across Development in Both Mouse and Human," *Proceedings of the National Academy of Sciences* 109, no. 40 (2012): 13–21.

212 **"Reckless sounds like you're not paying attention"** Dobbs, "Teenage Brains," 36.

212 **modern adolescence generates an awful lot of "weirdness"** Alison Gopnik, "What's Wrong with the Teenage Mind?" The Saturday Essay (blog), *Wall Street Journal,* January 28, 2012, available at: http://online.wsj.com/article/SB10001424052970203806504577181351486558984.html.

213 **the sheltered lives of modern adolescents** Margaret Mead, "The Young Adult," in *Values and Ideals of American Youth*, ed. Eli Ginzberg (New York: Columbia University Press, 1961), 37-51.

213 **"as-if" period** Rolf E. Muuss, *Theories of Adolescence*, 5th ed. (New York: McGraw-Hill, 1988), 72.

213 **"These Stone Age tendencies"** Jay Giedd, interview with Neal Conan, *Talk of the Nation*, NPR, September 20, 2011.

213 **"behavior that we would consider precocious"** Mintz, *Huck's Raft,* 68, 75, 87.

213 **But by the twentieth century** Ibid., 197.

214 **December 1924 issue of *Woman Citizen*** Zelizer, *Pricing the Priceless Child,* 67.

214 **It published an essay version** Margaret Sanger, "The Case for Birth Control," *Woman Citizen* 8 (February 23, 1924): 17–18.

214 **The divergent paths to American adulthood** Jeylan T. Mortimer, *Working and Growing Up in America* (Cambridge, MA: Harvard University Press, 2003), 9.

215 **the word "teenager" emerged in the American lexicon** Mintz, *Huck's Raft,* 239.

215 **"They live in a jolly world"** Quoted in ibid., 252.

215 **"Teens, for the first time, shared a common experience"** Ibid., 286.

215 **a portrait of high school culture in the Midwest** James S. Coleman, *The Adolescent Society: The Social Life of the Teenager and Its Impact on Education* (Westport, CT: Greenwood Press, 1981).

216 **"We know so much better what makes them tick"** Hulbert, *Raising America,* 280.

216 **"children's 'aspirational age' has risen"** Chudacoff, *Children at Play,* 217.

217 **"Myface"** Carrie Dann, Lauren Appelbaum, and Eman Varoqua, "Clinton's Speech at Rutgers," First Read (blog), NBCNEWS.com, April 20, 2007.

217 **"And it's *freaking people out*"** Clay Shirky, interview with the author, April 20, 2011.

218 **"fueled cravings for an easy life"** Mintz, *Huck's Raft,* 230.

218 **"If it were my task, Mr. Chairman"** Quoted in Kutner and Olson, *Grand Theft Childhood,* 50–51.

219 **"you explicitly have to open a communication channel"** Mimi Ito, interview with the author, May 24, 2012.

220 **"really accessible"** Ibid.

220 **"Not only that, but parents live in a society"** Clay Shirky, interview with the author, April 20, 2011.

220 **"Friending your child"** Ibid.

221 **"What's interesting about cell phones and Facebook"** Nancy Darling, interview with the author, March 29, 2011.

222 **"While we were growing up"** Clay Shirky, interview with the author, April 20, 2011.

223 **10.4 times per week** Barbara K. Hofer, *The iConnected Parent: Staying Close to Your Kids in College (and Beyond) While Letting Them Grow Up* (New York: Free Press, 2010), 16.

224 **rock-bottom lowest marks** Galinsky, *Ask the Children,* ch. 2.

225 **"Adults are not less excessive in their behavior"** Phillips, *On Balance,* 38.

225 **"an overcoming . . . a disciplining"** Phillips, *Going Sane,* 129.

225 **"The helplessness born of experience"** Phillips, *On Balance,* 38.

225 **they in fact rise and fall in tandem with adult problems** Mintz, *Huck's Raft,* 345.

226 **"mid-life rumination scale"** Steinberg, *Crossing Paths,* 151.

227 **they didn't want a second adolescence at all** Ibid., 152.

229 **all of us go through eight stages of development** For Erikson's seminal work on the stages of development, see Erik H. Erikson, *Identity and the Life Cycle* (New York: W. W. Norton, 1994).

229 **"generativity versus stagnation"** Ibid., 103.

229 **"integrity versus despair and disgust"** Ibid., 104.

230 **"It is the acceptance of one's"** Ibid.

230 **parents of twelve- to seventeen-year-olds** US Census Bureau, "America's Families and Living Arrangements: 2012," American Community Survey, Current Population Survey, table F1, available at: http://www.census.gov/hhes/families/data/cps2012.html (November 2012).

230 **risk of depression during perimenopause** Salynn Boyles, "Nearing Menopause? Depression a Risk," WebMD.com, available at: http://www.webmd.com/menopause/news/20060403/nearing-menopause-depression-risk (accessed April 22, 2013).

234 **"as antidepressants . . . more dependent on their children"** Phillips, *On Balance*, 98.

234 **"For a child growing up"** Phillips, *Going Sane*, 220.

chapter six

237 **"But I am telling only half the truth"** Mary Cantwell, *Manhattan, When I Was Young* (Boston: Houghton Mifflin, 1995), 155.

238 **"There's really *fun* stuff about raising kids"** Robin Simon, interview with the author, April 4, 2011.

239 **"When I think of the word 'parenting'"** Nancy Darling, "Why Parenting Isn't Fun," Thinking About Kids (blog), *Psychology Today* (July 18, 2010), available at: http://www.psychologytoday.com/blog/thinking-about-kids/201007/why-parenting-isn-t-fun.

239 **"Hanging out watching videos"** Ibid.

239 **"They involve just sitting back"** Ibid.

240 **positive psychology was the most popular course** Tara Parker-Pope, "Teaching Happiness, on the Web," Well (blog), *New York Times*, January 24, 2008, available at: http://well.blogs.nytimes.com/2008/01/24/teaching-happiness-on-the-web.

241 **"Few of the experiences of happiness"** Sissela Bok, *Exploring Happiness: From Aristotle to Brain Science* (New Haven, CT: Yale University Press, 2010), 103.

242 **the Grant study** George Vaillant, interview with the author, March 8, 2013.

242 **"Their lives were too human for science"** Quoted in Joshua Wolf Shenk, "What Makes Us Happy?" *The Atlantic* (June 2009): 36–53.

242 **When I first meet Vaillant in Boston** George Vaillant, interview with the author, March 23, 2011.

242 **"Joy is connection"** George Vaillant, *Spiritual Evolution: A Scientific Defense of Faith* (New York: Broadway Books, 2008), 124.

242 **"It's how Freud saw sex"** George Vaillant, interview with the author, March 23, 2011.

243 **"It's the difference between watching *Emmanuelle*"** Ibid.

243 **"Excitement, sexual ecstasy, and happiness"** Vaillant, *Spiritual Evolution*, 125.

244 **He's very fond of quoting William Blake's** Ibid., 119.

244 ***mono no aware*** Gopnik, *The Philosophical Baby*, 201.

244 **"We feed children"** Lewis, *The Four Loves*, 50.

244 **"Christmas eve, beautiful night"** Brené Brown, "The Price of Invulnerability," live TEDxKC talk on August 12, 2010, posted October 10, 2012, available at: http://tedxtalks.ted.com/video/TEDxKC-Bren-Brown-The-Price-of.

245 **She told the audience** Ibid.

245 **Brown calls this feeling** Ibid.

245 **"running around inside someone else's body"** Christopher Hitchens, *Hitch-22: A Memoir* (New York: Twelve Books, 2011), 338.

245 **"Joy is grief inside out"** Vaillant, *Spiritual Evolution,* 133.

248 **"Duty is one of those words"** John Lanchester, *Family Romance: A Love Story* (New York: Putnam, 2007), 154.

249 **"One is freed of the constant pressure"** Csikszentmihalyi, *Flow,* 179.

249 **"Set thy heart upon thy work"** Bhagavad Gita, trans. Juan Mascaro, rev. ed., (New York : Penguin Classics, 2003), 2:47.

250 **"I didn't have children because I wanted"** George Vaillant, interview with the author, March 23, 2011.

250 **"Here's what's coming to mind"** Ibid.

251 **"Suppose there were an experience machine"** Robert Nozick, *Anarchy, State, and Utopia* (New York: Basic Books, 1974), 42.

251 **"of profound connection with others"** Robert Nozick, *Examined Life: Philosophical Meditations* (New York: Simon & Schuster, 1990), 117.

251 ***eudaimonia*** See, for example, Sarah Broadie, "Aristotle and Contemporary Ethics," in *The Blackwell Guide to Aristotle's* Nicomachean Ethics, ed. Richard Kraut (Malden, MA: Blackwell, 2006), 342.

251 **"Children are a reason to get up in the morning"** Robin Simon, interview with the author, April 4, 2011.

251 **parents are much less likely to commit suicide** Émile Durkheim, *Suicide: A Study in Sociology,* ed. George Simpson, trans. John A. Spaulding and George Simpson (New York: Free Press, 1979), 197–98.

252 **"anomie" and "normlessness"** Ibid., 241 et seq.

252 **"In an anomic society, people can do as they please"** Jonathan Haidt, *The Happiness Hypothesis: Finding Modern Truth in Ancient Wisdom* (New York: Basic Books, 2006), 175.

252 **"The love we feel for children"** Gopnik, *The Philosophical Baby,* 241.

253 **parents who have custody of their children** Ranae J. Evenson and Robin W. Simon, "Clarifying the Relationship Between Parenthood and Depression," *Journal of Health and Social Behavior* 46 (December 2005): 355.

253 **depression surveys often ask questions** See, for example, the Center for Epidemiologic Studies Depression Scale (CES-D), a copy of which can be found in Center for Substance Abuse Treatment, *Managing Depressive Symptoms in Substance Abuse Clients During Early Recovery: Appendix B* (Rockville, MD: Substance Abuse and Mental Health Services Administration, 2008), available at: http://www.ncbi.nlm.nih.gov/books/NBK64056.

253 **"For many, the lack of structure of those hours"** Csikszentmihalyi, *Flow,* 168.

254 **"Sunday neurosis"** Viktor Frankl, *Man's Search for Meaning* (Boston: Beacon Press, 1992), 112.

254 **"If architects want to strengthen a decrepit arch"** Ibid., 110.

254 **"a zest for life in all its complexity"** Bok, *Exploring Happiness,* 117.

254 **"best piece of poetrie"** Ben Jonson, "On My First Sonne" (c. 1603).

255 **the "experiencing self" versus the "remembering self"** Kahneman, *Thinking, Fast and Slow,* 381.

256 **can be deformed by a bad ending** Ibid.

256 **in a 2010 TED lecture** Daniel Kahneman, "The Riddle of Experience vs. Memory," TED Talk, February 2010, posted March 2010, available at: http://www.ted.com/talks/daniel_kahneman_the_riddle_of_experience_vs_memory.html.

257 **85 percent of all parents** Pew Research Center, "As Marriage and Parenthood Drift Apart, Public Is Concerned About Social Impact," July 1, 2007, available at: http://www.pewsocialtrends.org/files/2007/07/Pew-Marriage-report-6-28-for-web-display.pdf.

257 **"Especially things like reading books to them"** Mihaly Csikszentmihalyi, interview with the author, July 25, 2011.

258 **"In our interviews, there's a section"** Dan P. McAdams, interview with the author, January 8, 2013.

258 **Storytelling . . . is our natural response** Kahneman, "The Riddle of Experience vs. Memory."

258 **remembering selves are in fact *who* we are** Kahneman, *Thinking, Fast and Slow,* 390.

258 **"You don't have a good story"** Dan P. McAdams, interview with the author, January 8, 2013.

258 **"I think this boils down to a philosophical question"** Quoted in Jennifer Senior, "All Joy and No Fun: Why Parents Hate Parenting," *New York,* July 4, 2010.

259 **"Highly generative adults . . . invest considerable time"** Dan P. McAdams, "The Redemptive Self: Generativity and the Stories Americans Live By," *Research in Human Development* 3 (2006): 93.

259 **he frequently hears from fathers** Dan P. McAdams, interview with the author, January 8, 2013.

259 **"the redemptive stories our mothers tell"** Kathryn Edin and Maria Kefalas, *Promises I Can Keep: Why Poor Women Put Motherhood Before Marriage* (Berkeley: University of California Press, 2005), 11.

260 **"bring different worlds to your home"** Philip Cowan, interviews with the author, February 2, 2011, and March 10, 2011.

261 **"It comes out as, 'I've developed a story'"** Dan P. McAdams, interview with the author, January 8, 2013.

261 **"the evaluator shouldn't be the past generation"** Ibid.

261 **"Noncalorie chocolate"** Dan Gilbert, interview with the author, March 22, 2011.

261 **the helplessness of modern American parents** See, especially, Mead, *And Keep Your Powder Dry.*

263 **"Having found myself faced with that old bull-session question"** Marjorie Williams, "Hit by Lightning: A Cancer Memoir," in *The Woman at the Washington Zoo,* ed. Timothy Noah (New York: PublicAffairs, 2005), 321.

index

Aassve, Arnstein, 82
abduction paranoia, 168–69
adolescence/adolescents
brain of, 208–12, 218
conflict and, 195–98, 207
as critical of parents, 194
culture and, 188, 212, 215–16, 238
dependency of, 212–18
"discovery" of, 187–88
"emerging adulthood" and, 214–15
excesses and, 223–26
independence and, 209, 211, 217, 218, 219
marital relations and, 48, 73, 201–8
paradox of modern, 216–17, 223
parents' identification/experience with, 186–89, 224–33
personal preferences of, 196–98
power issues and, 193, 195, 220–23
privacy of, 189, 221–22
protection and support of, 12, 188–89, 213, 216, 217, 234–35, 238
rating of parents by, 194
regrets and, 226–33
relationships between same sex parent and, 198–99
as repeat of toddler years, 192–93, 195
risk and, 209–10, 212, 214
schools and, 214
surveillance/creeping on, 221–22
technology and, 216, 217, 218–23
transition of parents during, 185–90, 195–96
See also specific topic
allowances, 128
Amato, Paul, 54
American Time Use Survey, 53, 122, 158
anxiety
boredom and, 32
child safety and, 168
deadlines and divided time and, 59–60
flow and, 32
and elusive goals of parenting, 138
legacies and, 262
overscheduled parents and, 123

as-if period, 213, 214–15
attachment parenting, 111, 136, 151, 242–43. *See also* bonding; connection
attention: flow and divided, 34–40
authority, parental, 129, 192
autonomy
  adolescence and, 196, 209
  and balancing professional and family obligations, 40–44
  and definition of liberty, 43
  early years of parenting and, 17–18
  flow and, 28–34, 237
  as freedom from obligations, 249
  and professional ambitions, 40–44
  sleep needs and, 20–23
  women's movement and, 42–43

baby products, 154
Baumeister, Roy F., 22–23
Bettelheim, Bruno, 216
Beyoncé, 197–98
Bianchi, Suzanne, 158–59
Blake, William, 102, 244
blogs, parenting, 189
Bok, Sissela, 241, 254
Bombeck, Erma, 9, 17
bonding, 111, 112, 243, 252, 253, 265. *See also* attachment parenting; connection
Bonnet, Michael H., 22
boredom, 32–33, 34, 166–67
Borgmann, Albert, 105
Boy Scouts, 129–30, 170–71
Bradbury, Thomas, 203–4
brain, 26–28, 106, 208–12, 218
Brazelton, T. Berry, 151, 175
Bronfenbrenner, Urie, 69
Brown, Brené, 198, 244–45
Burgard, Sarah A., 55
Cantwell, Mary, 237
Casey, B.J., 209–10, 211
cell phones, 219, 221, 223
Cherlin, Andrew, 42, 43
child care, 57, 58, 67, 81, 82, 89–90, 122, 149–50, 151, 161–62. *See also* household labor
Child Care Aware of America, 83
child labor, 9, 127–28, 214
children
  in the age of globalization, 131–44
  as always changing, 106–10
  balancing among several, 172–73
  decline in number of, 122
  as "economically worthless but emotionally priceless," 126–31
  as high-cost/high-reward, 6, 12–13, 143
  historical views of, 126–28, 213–14
  indoor, 162–71
  isolation of, 165
  as living in permanent present, 27–28, 100
  overscheduled, 119–26, 165–66
  protection and support of, 9–10, 12, 167–68, 188–89, 213, 216, 217, 234–35, 238
  school as primary job of, 128
  self-esteem of, 176
  sentimentalization of, 164, 167
  sexual abuse of, 168
  as superegos, 261
  as vulnerable and innocent, 127
  See also specific topic
Christensen, Andrew, 204, 205–6
Chua, Amy, 133, 147, 176–77
Chudacoff, Howard, 164, 216
college
  admissions to, 133
  as preparation for motherhood, 152

college-educated women, 19, 144
community. *See* neighbors/ neighborhoods
concerted cultivation
  in the age of globalization, 131–44
  child safety and, 122, 167–69, 238
  and creating happy children, 171–77
  definition of, 120
  electronic media and, 122, 169–71, 238
  homework and, 177–81
  indoor children and, 162–71
  overscheduled parents and, 119–26
  overview about, 237–38
  pressures on mothers and, 144–59
conflict
  adolescence and, 195–98, 207
  in marriage, 49–50, 54
  between mothers and daughters, 198
Conley, Dalton, 37, 38
connections, 201, 242–44, 245, 251, 253, 263. *See also* attachment parenting; bonding
"consumer parenting," 91
Coontz, Stephanie, 42, 43, 66
Correll, Shelley, 143
Cowan, Carolyn, 48–49, 54, 56, 60, 85, 92
Cowan, Philip, 48–49, 54, 56, 85, 89, 92, 259, 260
Crawford, Matthew B., 104, 105, 106
"cry-it-out" method, 86–88, 136
Csikszentmihalyi, Mihaly, 30–32, 33–34, 107, 249, 253–54, 257
Cub Scouts, 117–19, 129–30, 177, 179
Cummings, E. Mark, 49
Cunningham, Michael, 79
Czerwinski, Mary, 37
Darling, Nancy, 166–67, 196–97, 207, 210, 221, 239, 259, 260
dating, adolescent, 205
Davila, Joanne, 193
day care, 83, 168
Deaton, Angus, 6
Dederer, Claire, 155
depression, 230, 253, 254, 261
Descartes, René, 108
Dinges, David, 21
discipline issues, 10, 67–71, 207
division of labor. *See* household labor
divorce, 49, 158, 198, 205, 232
Dobbs, David, 212
Doherty, William, 6, 12, 72–73, 91, 119, 238
domestic scientists, women as, 153–54
Druckerman, Pamela, 69–70, 91
Duke University Talent Identification Program, 134
duty: parenting as, 247–54

Early Childhood Family Education (ECFE), 15–17, 19
Edin, Kathryn, 259–60
"ego depletion," 22
electronic media. *See* technology
"emerging adulthood," 214–15
Emory University: obedience study at, 69
Ephron, Nora, 137–38, 185
Erikson, Erik, 229, 230, 235, 259, 260
excess
  adolescence and, 223–26
  young children and, 23–28
"experience machine," 250–51
Experience Sampling Method (ESM), 30–31, 33–34
"experiencing self," 255–56, 258

Facebook, 217, 220–21
fairness, 54–55, 56, 57, 79, 88, 89, 234
Families and Work Institute, 159–60, 181
family planning, 7–8
fathers
  adolescence and, 205, 207
  and "autumnal" relationship with sons, 136
  and change in experience of parenting, 8, 9
  concerted cultivation and, 122, 149
  discipline responsibilities of, 207
  divided time and, 58
  and division of household labor, 53
  happiness of, 4, 5, 71
  and how much time parents spend with children, 181
  narratives of, 259
  obedience issues and, 70–71
  social isolation of, 63–64
Fink, Kathryn, 61
flow
  and absence of feeling, 30
  autonomy and, 28–34, 237
  boredom and, 32–33, 34
  characteristics of, 31
  definition of, 30
  happiness and, 31, 34
  measurement of, 30–31
  multitasking and, 35–36, 37, 38
  rules and, 31, 33, 34
  solitude and, 33
  work and, 33–40
France, 83, 91–92
Frankl, Viktor, 254
freedom; *See* autonomy
Freud, Sigmund, 102, 242, 261
Friedan, Betty, 152–53, 154
Friedman, Tom, 138–39

Gagliardi, Annette, 31, 65
Galinsky, Ellen, 160, 181, 194, 224
Giedd, Jay, 213
Giffords, Gabrielle, 156
gift-love, 111–12, 113
"gift of service," 180
Gilbert, Daniel, 27, 32–33, 71, 261
Gilovich, Tom, 258
globalization and children, 131–44
Gopnik, Alison, 26, 31, 101, 106–7, 112, 114, 212, 244, 249, 252–53
government: role in parenting of, 81–83, 92
Grant Study, 242
Groves, Ernest, 174–75
guilt, 38, 75, 78, 86, 87–88, 89, 90–91, 168, 264

Haidt, Jonathan, 252
Hall, Stanley, 187, 188
happiness
  burden of, 171–77
  as by-product, 251
  complexity of studies about parenting and, 157–58
  definition of, 240–41, 243, 254
  and division of household labor, 57
  as doing, 251, 254
  duty and, 250
  and early years of parenting, 17–18
  expectations of, 248
  and "experience machine" experiment, 251
  fairness and, 234
  flow and, 31, 34
  fun differentiated from, 238–39

and gap between fathers and mothers, 71
and goals of parenting, 234, 238, 251
and government role in parenting, 81–83
marriage and, 57, 72, 203
measuring, 241
of parents, 4–6, 48, 158, 194, 238, 253
remembering self and, 256–57, 258
Simon's studies of parenting and, 238–39, 253
of single parents, 48, 158, 253
sleep and, 21
working mothers and, 5
Harvard University, 10, 27, 69
"having it all," 41
Hays, Sharon, 151, 155, 175, 245, 246, 249, 261, 262, 263, 264, 265
Head Start, 69
Hitchens, Christopher, 245
Hochschild, Arlie Russell, 53–54, 89, 158
Hofer, Barbara, 223
Holt, Evan, 54
Holt, Nancy, 54
homework, 128, 177–81
household labor
and change in experience of parenting, 8
division of, 8, 50–57, 80–81, 237
and women as domestic scientists, 153–54
Hulbert, Ann, 151–52, 174
hyperparenting, 119–26

Institute for American Values, 48
Internet. *See* technology
Ito, Mimi, 171, 219, 220

Jong, Erica, 155
joy
as connection, 242–44, 245
fear and, 244–45
"forboding," 245

Kagan, Jerome, 7–8, 176, 180
Kahneman, Daniel, 5, 21, 83, 255, 256, 258
Kaiser Family Foundation, 170
Karolyi, Bela, 121
Kaufman, Sue, 155
Kefalas, Maria, 259–60
Kettering Medical Center (Dayton), 22
Killingsworth, Matthew, 5–6
kindergartners: obedience of, 69
Kolod, Todd, 28, 71, 74
Konner, Melvin, 15
Kuhn, Peter, 144
Kumon (after-school enrichment program), 146
Kundera, Milan, 98

Lanchester, John, 248
Lareau, Annette, 119–20, 124, 129, 144, 149–50, 166, 167, 181
Leach, Penelope, 175
LeMasters, E.E., 47
lesbian couples, 49, 60
Lewis, C.S., 99, 111, 113, 244, 245
Lewis, Michael, 91
life-redemption narratives, 259–60
"little adult" problem, 28
Louis C.K., 9
love
attachment and, 242–43
caring and, 112, 249
duty and, 248, 250
gift-, 111–12, 113, 244, 265

love (*continued*)
need-, 111–12
and stages of adulthood, 229
Lozano, Fernando, 144

Mansbach, Adam, 55–56
marriage
adolescence and, 48, 201–8
age at first, 42
conflict in, 49–50, 54
and division of household labor, 50–57, 237
expectations for, 41–44, 72
fairness and, 54–55, 56, 57, 79, 88, 89
gratitude in, 89
happiness/satisfaction in, 47–49, 57, 72, 203, 205
as public institution, 71–72
redefining attitudes in, 93
sex and, 50, 71–76, 203
social isolation and, 61–67
"soul mates" and, 72
Martin, Edward Sandford, 117
Matthews, Gareth B., 59–60, 108, 109–10
Mattingly, Marybeth, 59
McAdams, Dan P., 258, 259, 260–61
McHale, Susan, 205
Mead, Margaret, 134–36, 137, 138, 139, 176, 213, 261–62
Meetup, 61
Meyer, David E., 38
midlife rumination scale, 226–27
midlife crisis, 187, 207–8
Mintz, Steven, 126–27, 137, 165, 168, 213, 215, 218, 225
Missouri City, Texas, 162–63, 169, 171–72.
"mommy mystique," 159
Montgomery-Downs, Hawley, 22
Motherlode (blog), 175
mothers
adolescence and, 207
concerted cultivation and, 122
daughters' conflict with, 198
discipline responsibilities of, 207
empty nest, 5
happiness of, 4–5, 71
high standards for, 144–59
and how much time parents spend with children, 181
and "intensive mothering," 155
as nags, 68, 150
"pressure points" of, 150
as primary caretakers, 149–50
social isolation of, 61–64
wage gap between childless women and, 143
multitasking, 3, 35–36, 37, 38, 58, 159

narratives: identity, 259–61
National Institute of Mental Health, 213
National Marriage Project, 19
National Vital Statistics, 19
need-love, 111–12
neighbors/neighborhoods, 64–66, 160–61, 252
nocturnal caregiving, 55–56, 162
normlessness, 7, 252
Nozick, Robert, 250–51

Obama, Barack, 45
Ondaatje, Michael, 95
overscheduled parents, 119–26, 167

parents/parenting
children's rating of, 194

cultivation as primary obligation of, 135–38
deferring of, 19
expectations about, 11, 18, 150–51, 250
goals of, 10, 135, 138, 175, 176, 234, 238, 241
government role in, 81–83, 92
happiness of, 4–6, 48, 158, 194, 238, 253
as "high-cost/high-reward activity," 6, 12–13, 143
midlife reflections of, 226–27
midlife crisis of, 187
overscheduled, 119–26, 167
and "parent" as a verb, 152, 238
regrets of, 231–33
Paul, Pamela, 165
permanent present: children as living in, 27–28, 100
Perry, Katy, 101
"pervasive busyness" sensation, 65
Phillips, Adam, 18, 24–26, 41, 44, 75, 102–3, 175, 193–94, 224, 225, 234
philosophy, 106–10
play, 28–29, 77–78, 101, 164–65, 167, 262
playgrounds, 167–68
playrooms, 163–65
Popenoe, David, 72
pornography: affects on adolescents, 211
positive psychology, 240
prefrontal cortex, 26–27, 31, 100, 101, 210, 211
preschoolers. *See* toddlers/preschoolers
Putnam, Robert, 62, 64–65

remembering self, 254–58
risk: adolescence and, 209–10, 212, 214
Roberts, J.M., 18
Rosin, Hanna, 53
Rossi, Alice, 3–4
Rousseau, Jean-Jacques, 127

safety, child, 122, 167–69, 188, 196, 197, 238
Sandberg, Sheryl, 40–41
sandwich generation, 67
Sayer, Liana, 59–60
"schmoozers," 62
self-control/restraint, 22–23, 25, 49
sex
adolescence and, 183–84, 199
marriage and, 50, 71–76, 203
Shapiro, Dani, 183
Shirky, Clay, 218, 220–21, 222
Simon, Robin, 82, 238–39, 251, 252, 253
single parents, 5, 48, 82, 155–59, 253, 259–60
Skinner, B.F., 36, 242
Slaughter, Anne-Marie, 41
sleep, 6, 20–23, 55–56, 75
sleep-training, 86–88, 89
social networks/interactions, 61–67, 217, 219
social safety nets, 81
Spock, Benjamin, 32, 62–63, 175, 176, 234
stages of adulthood, 229–30
stay-at-home parents, 18, 36, 55, 63–64, 154, 155, 199, 227–28
Steinberg, Laurence, 186–87, 189–90, 195, 198, 199, 205, 207–8, 211, 226–27
Steinberg, Wendy, 189
Stone, Arthur, 6
Stone, Linda, 37
stress, 37, 70–71, 82, 90, 189, 204

Sugar Land (Houston suburb), 131–34, 169
suicide, 251–52
summer camp, 121–22
Suzuki method, 179

task-switching, 37–38
technology, 122, 169–71, 216, 217, 218–23, 238
Tierney, John, 22–23
Tiger Moms, 93, 130, 133, 139, 142, 145–46, 147
time
  decline in adolescent-family, 193
  decline in couple's together, 72–73
  divided, 58–61
  efficiency of, 79
  gender differences in experiencing of, 79–80
  and how much time parents spend with children, 181
toddlers/preschoolers, 25, 26, 68, 69, 73, 106, 192–93, 195
"Top 10 Percent Rule," Texas, 133, 141
toys, 164, 165

University of California, Los Angeles: dual-earner study at, 55
University of Georgia: obedience study at, 69
University of Michigan: adolescent study at, 207

Vaillant, George, 242–44, 245, 250

Waldman, Ayelet, 155
Warner, Judith, 12, 82–83, 92, 143, 159
Weininger, Elliot, 149–50
Wertham, Fredric, 218
White House Middle Class Task Force, 143
Whitehead, Barbara Dafoe, 72
Whyte, William H., 128–29
Williams, Marjorie, 263, 264–65
Winnicott, Donald, 103
women
  and balancing professional and family obligations, 42–43
  college dropout rate of, 42–43
  as domestic scientists, 153–54
  as "having it all," 41
  midlife crisis of, 207–8
  in the 1950s, 152–53
  and wage gap between childless women and mothers, 143
  See also mothers
women's movement, 42–43, 152–53
work
  and change in experience of parenting, 8–9
  and deferring professional ambitions, 40–44
  flow and, 33–40
  fulfillment of parents with, 199, 201
  portability and accessibility of, 35, 37
  and division of household labor, 53, 55
  flow and, 36
  happiness and, 5

Zelizer, Viviana, 9–10, 128, 170, 174, 187

about the author

JENNIFER SENIOR is a contributing editor at *New York* magazine. She lives in New York with her family.